DESIGN of THERMAL OXIDATION SYSTEMS for VOLATILE ORGANIC COMPOUNDS

DESIGN of THERMAL OXIDATION SYSTEMS for VOLATILE ORGANIC COMPOUNDS

David A. Lewandowski

LEWIS PUBLISHERS
Boca Raton London New York Washington, D.C.

Library of Congress Cataloging-in-Publication Data

Lewandowski, David A.
    Design of thermal oxidation systems for volatile organic compounds
/ by David A. Lewandowski.
        p.   cm.
    Includes bibliographical references and index.
    ISBN 1-56670-410-3 (alk. paper)
    1. Volatile organic compounds—Environmental aspects. 2. Waste
gases—Purification. 3. Oxidation. I. Title.
TD885.5.O74L48 1999
628.5'32—dc21   99-35424
                                                                CIP

No claim to original U.S. Government works
International Standard Book Number 1-56670-410-3
Library of Congress Card Number 99-35424
Printed in the United States of America 1 2 3 4 5 6 7 8 9 0
Printed on acid-free paper

Preface

Air pollution's origin can be traced to the discovery of fire. Sub-micron particles in the form of smoke were emitted from incomplete combustion. However, air pollution did not become a real problem until the industrial revolution. Today, over two billion pounds of pollutants are released to the atmosphere each year. This book focuses on one technology, thermal oxidation, which can be used to reduce the emissions of volatile organic compounds (VOCs) to the atmosphere.

The term volatile refers to the tendency of a compound to evaporate. By definition, volatile organic compounds are "organic." While there are more than 100 natural and man-made chemical elements in total, organic compounds consist only of the elements carbon (C), hydrogen (H), nitrogen (N), oxygen (O), sulfur (S), and chlorine (Cl). VOCs are emitted from a myriad of industrial processes. As a consequence of the 1990 Clean Air Act Amendments, more and more industries are being required to reduce their VOC emissions. These include the chemicals, petro-chemicals, coatings, textiles, rubber, pulp and papers, metals, pharmaceuticals, food, and minerals industries.

The control of emissions of volatile organic compounds has been a very prominent environmental issue in the 1990s and will continue into the 21st century. No single technology has played as important a role in the control of VOC emissions as thermal oxidation. Its prominence is a result of its ability to destroy VOCs in a one-step process while producing innocuous by-products. Control of VOC emissions will increasingly rely on thermal, regenerative, and catalytic oxidizers as environmental regulations become more stringent. The market for this equipment is projected to exceed $2.1 billion in the year 2000.

Thermal oxidation is a combustion process. Combustion is defined as the burning of gases, liquids, or solids, in which the substance is oxidized evolving heat and often light. Thermal oxidizers are distinguished from incinerators by virtue of the fact that they generally only treat vapors or relatively pure liquids. Thus, the combustion products usually do not contain particulates, dioxins, or heavy metals, the species that have made incinerators a somewhat controversial treatment device. Many books available in the literature focus on hazardous wastes. VOCs are not normally classified as hazardous waste and thus are not subject to the more stringent regulations and design requirements imposed on hazardous waste incinerators.

This book is intended as a resource of information necessary to develop conceptual designs of a thermal oxidation system. It is primarily intended for those who are well versed in engineering but are unfamiliar with thermal oxidation technology. While most engineers possess knowledge of the engineering fundamentals necessary to understand the design of thermal oxidation systems, they need a reference to lead them through this design. They may not refer to this book because they will be designing and fabricating thermal oxidizers themselves, but because plants in which they are working must install these systems. They must be able to intelligently develop equipment specifications for thermal oxidation systems and evaluate bids when received. The same can be said of engineers working for Architect/Engineers (AE). They must have a firm grasp of thermal oxidizer design principles so that they can prepare equipment specifications and evaluate bids.

This book progresses from fundamental concepts to more advanced features. It not only includes design principles for basic thermal oxidizers, but also covers regenerative and catalytic oxidizers. Concise information on heat recovery systems, the latest technology for pre- and post-combustion NOx control, and appendices containing physical and chemical property data for hundreds of VOCs are also included.

Others technologies are available for treatment of VOCs. However, most require further treatment of the VOCs once they are removed from the gas stream. Thermal oxidizers are favored in most circumstances because of their reliability, the fact that usually no further treatment is required, and because of their ability to achieve high VOC destruction efficiencies. The prevalence of combustion systems such as boilers, furnaces, fired heaters, and burners in a myriad of manufacturing and production facilities has aided the acceptance of thermal oxidizers by operators.

About the Author

David Lewandowski's academic background includes a Bachelor's degree in Chemistry from Penn State University and a Master's degree in Chemical Engineering from Cleveland State University. He is also a registered Professional Engineer in the State of Pennsylvania. He possesses over twenty-five years of industrial experience at Diamond Shamrock Corporation, Westinghouse, Process Combustion Corporation, and Consol Energy. He has also performed independent consulting.

Mr. Lewandowski's career has focused on process engineering, in general, and combustion and other high temperature systems, in particular. This combustion experience encompasses coal, gas, oil, and both hazardous and nonhazardous wastes. His work has encompassed process design, process development, testing, research, and computer modeling.

He holds one patent on a system to reduce nitrogen oxide emissions from thermal oxidation systems and another on a purge system for a Regenerative Thermal Oxidizer (RTO). He has presented papers at numerous conferences and is a member of the American Institute of Chemical Engineers (AICHE) and the Air & Waste Management Association (AWMA).

Acknowledgments

The author would like to acknowledge the following individuals who provided invaluable review and comments during the production of this book:

Gene McGill, Gene McGill & Associates
Dan Banks, Banks Engineering
Greg Homoki, Alstom Energy Systems Inc.
Jack Rentz, Rentech Boiler Systems Inc.
Joseph Bruno, AirPol Inc.
Michael DeLucia, Harbison-Walker Refractories Co.
Hassan Niknafs, Norton Chemical Process Products Corp.
Edward Donley, Process Combustion Corp.

Contents

Chapter 5 Mass and Energy Balances

Chapter 6 Waste Characterization and Classification

Chapter 7 Thermal Oxidizer Design

Chapter 11 Combustion NOx Control

Chapter 12 Postcombustion NOx Control

Chapter 13 Gas Scrubbing Systems

Chapter 14 Safety Systems

Chapter 15 Design Checklist

1 Introduction

CONTENTS

The control of emissions of volatile organic compounds (VOC) became a very prominent environmental issue with the passage of the 1990 Clean Air Act Amendments (CAA). While environmental regulations were in existence long before that time, they focused on controlling the concentrations of six "priority pollutants" in the ambient air. These pollutant species were ozone, nitrogen oxides, carbon monoxide, lead, particulates, and sulfur dioxide. The 1990 CAA amendments changed the focus to control of emissions of a set of specific chemical compounds called volatile organic compounds (VOC).

No single technology has played as important a role in the control of VOC emissions as thermal oxidation. Its prominence is a result of its ability to destroy VOCs in a one-step process while producing innocuous by-products (for the most part). VOCs are not normally classified as hazardous waste and thus thermal oxidizers are not normally subject to the more stringent design and operating requirements imposed on hazardous waste incinerators.

1.1 COMBUSTION

Thermal oxidation is a combustion process. Combustion is defined as the burning of gases, liquids, or solids, in which the substance is oxidized, evolving heat and often light. Thermal oxidizers are distinguished from incinerators by virtue of the fact that they generally only treat vapors or relatively pure liquids. Thus, the combustion products usually do not contain particulates, dioxins, or heavy metals, the species that have made incinerators a somewhat controversial treatment device. As described in this book, the term "thermal oxidation" is synonymous with "afterburner."

Other technologies are available for treatment of volatile organic compounds (VOC). However, thermal oxidizers are favored in most circumstances because of their reliability, the fact that usually no further treatment is required (not always true, as will be explained in Chapter 13), and because of their ability to achieve high VOC destruction efficiencies. The prevalence of combustion systems such as boilers, furnaces, fired heaters, and burners in a myriad of manufacturing and production facilities has aided the acceptance of thermal oxidizers by operators. Most other treatment technologies require further treatment of the VOCs once they are removed from the gas stream. An

1

example is carbon adsorption. Here, the waste gas containing the VOC is forced through a bed of activated carbon. The VOCs are adsorbed into the pores of the carbon and clean gas exhausts from the opposite end. The drawback lies in the fact that the VOC has not been destroyed but transferred from the waste gas to the carbon adsorbent. The carbon bed must be regenerated. This can either be done on-site or by interchanging spent carbon canisters with fresh canisters. While feasible and a common practice in many applications, it adds to the complexity of the operation. Again, thermal oxidation is generally a one-step process obviating the need for further processing.

1.2 HISTORY OF AIR POLLUTION

While regulation of air emissions has been a relatively recent trend in contemporary history, the generation of noxious fumes from industrial processes has been around for hundreds of years. In the 14th century, a man was tortured in England for emitting a "pestilent odor" to the atmosphere from burning coal. Fumes from coal burning became so bad in the late 1550s that the English parliament passed a law forbidding the burning of coal in London when parliament was in session. The first air pollution episodes in the U.S. were also from burning coal. Chicago adopted a smoke control ordinance in 1881. In those days, the responsibility for controlling air emissions rested with state and city governments. However, pollution controls laws did not become prevalent in the U.S. until after World War II. These were usually enacted after the effects of air pollution were undeniable.

One of the first recorded public health episodes resulting from air pollution occurred in the small industrial town of Donora, PA in 1948. Steel, zinc, and sulfuric acid plants poured noxious gases into the air on a daily basis. Topographically, Donora is located in a valley. Usually winds dispersed these gases over great distances. However, in a 5-day period between October 26th and 30th, 1948, the noxious gases accumulated in the valley to such an extent that thousands of people became ill and 20 died.

A similar incident occurred in Poza Rica, Mexico in 1950. A natural gas recovery and "sweetening" process was starting up. Sweetening consists of removing the hydrogen sulfide that is a normal constituent of natural gas. Due to an equipment malfunction, a large quantity of hydrogen sulfide escaped into the atmosphere at the same time that a thermal inversion enveloped the area. There were 320 people hospitalized and 22 deaths. A thermal inversion also caused a buildup of noxious gases in London in December of 1952. At least 4000 deaths were attributed to the polluted air.

Industrial sources were not the only source of air pollutants. The popularity of the automobile produced more air emissions than industrial sources in some areas. One of these areas was Los Angeles, CA. Health effects stemming from air contaminants began to appear in Los Angeles after World War II. Industry was initially blamed for producing a brown haze that became known as smog. It consisted of a mixture of nitrogen oxides (NOx), sulfuric acid mist from the condensation of sulfur dioxide, and particulates. But even after emissions from industrial sources were reduced, the smog persisted. Eye and skin irritation and plant damage could not be attributed to industrial pollutants alone. It was then discovered that automobiles were a major factor in the generation of this smog.

1.3 THERMAL OXIDATION'S WIDE APPLICABILITY

The first major national legislation enacted to control air emissions was the Clean Air Act, passed in 1955. It has been amended several times since. However, the most dramatic amendments were added in 1990. The resulting regulations serve as the basis for reduction of millions of pounds of VOCs emissions to the atmosphere annually. The most effective technology for reaching the mandated emission reduction is thermal oxidation.

By definition, VOCs are "organic." While there are more than 100 natural and man-made chemical elements in total, organic compounds consist only of the elements carbon (C), hydrogen (H), nitrogen (N), oxygen (O), sulfur (S), and chlorine (Cl). Other elements are sometimes present, but over 99% of the organic compounds in existence are comprised of these six elements. However, not all organic compounds are volatile. "Volatile" refers to the tendency of a compound to evaporate. In many cases, compounds classified as VOCs are gases at standard temperature and pressure. Solid organic compounds are usually not volatile. An example of an organic compound that is not volatile is glucose (sugar). Generally, organic compounds with high molecular weights (>200) are not volatile. Neither are viscous liquids.

Not all air emissions can be treated using thermal oxidation. Inorganic particulate emissions are one category for which thermal oxidation is ineffective. This is unfortunate since emissions of fine particulate matter (< 2.5 microns in size) has been identified as a major health concern. Nor can thermal oxidation be used to treat emissions from nonindustrial sources. Automobiles fit this category. While we normally think of automobiles as emitting carbon monoxide and nitrogen oxides, they are also a source of VOC emissions. Table 1.1 shows VOC emissions from passenger cars tested in 1995.

TABLE 1.1
Passenger Car Tail-Pipe Emissions
(grams emitted per kilometer driven)

Organic Compound Emitted	Car #1	Car #2	Car #3
Benzene	0.0183	0.0184	0.0123
Heptene	0.0021	0.0016	0.0014
Heptane	0.0125	0.0058	0.0077
Toluene	0.0301	0.0289	0.0178
Ethylbenzene	0.0066	0.0075	0.0027
m/p-Xylene	0.0130	0.0140	0.0054
o-Xylene	0.0054	0.0042	0.0023
Isopropyl benzene	0.0003	0.0004	0.0002

Source: Journal of the Air & Waste Management Association, Volume 45, February 1995.

Values shown represent grams emitted per kilometer driven. A comparison is made between three different car models that include catalytic converters.

Emission of air pollutants is certainly not a phenomenon of the U.S. alone. However, on a per capita basis, emissions in the U.S. generally exceed the emissions from European countries and Japan. When compared on a per capita basis, U.S. emissions are approximately 15 times greater than Japanese emissions and 10 times greater than West German and Swedish emissions.

1.4 AIR POLLUTANT EMISSIONS IN THE UNITED STATES

All industrial sources in the U.S. must report their annual air emissions to the US Environmental Protection Agency (EPA). The EPA compiles these data in a database called the Toxics Release Inventory (TRI). In 1993, a total of 23,000 facilities reported emissions data to the EPA. In total, approximately 2.8 billion pounds of toxics substances were released to the atmosphere that year. The top ten chemicals released and the quantities released are shown in Table 1.2. Of these ten, thermal oxidation technology could be used to treat all but hydrochloric acid and chlorine. Releases by industry category are shown for 1994 in Table 1.3.

TABLE 1.2
Top 10 Substances Released to the Air (1993)

Substance	Quantity Released (lb)
Toluene	177,301,267
Methanol	172,292, 981
Ammonia	138,057,165
Acetone	125,152,462
Xylene	111,189,613
Carbon disulfide	93,307,339
Methyl ethyl ketone	84,814,923
Hydrochloric acid	79,073, 655
Chlorine	75,410,108
Dichloromethane	64,313,211

Source: *Pollution Engineering*, February 1996.

Control of VOC emissions will increasingly rely on thermal, regenerative, and catalytic oxidizers as environmental regulations become more stringent. The market for this equipment is projected to exceed $2.1 billion by the year 2000.[1] It is estimated that 25% of these orders will be placed by the chemicals industry. The forest products and electronics industries are also expected to account for much of the demand.

Even though air emissions releases have declined in the U.S. from 3.75 billion pounds in 1988 to 2.168 billion pounds in 1995, the demand for thermal oxidation systems is growing, due to increasingly stringent environmental regulations.

TABLE 1.3
Air Emission Releases by Industry Category

Industry	Quantity of Air Emissions (lb)
Chemicals	884,903
Primary metals	483,224
Paper	248,976
Plastics	130,937
Transportation equipment	130,834
Fabricated metals	106,680
Petroleum refining	69,849
Furniture	50,881
Electrical	42,140
Lumber	33,516
Others	375,840
Total	2,557,780

Source: 1994 *EPA Toxic Release Inventory.*

1.5 INDUSTRIAL SOURCES OF AIR POLLUTION

VOCs are generated by a wide variety of industries in a wide variety of operations. A sampling of those industries is shown in Table 1.4. This abbreviated list demonstrates that the need for thermal oxidation systems is almost universal. Thus, a working knowledge of their design and operation is beneficial to mechanical engineers, chemical engineers, process engineers, environmental engineers, environmental regulators, and environmentalists.

TABLE 1.4
Industries Requiring Thermal Oxidation Systems

Chemicals	Oil
Organic	Production
Inorganic	Refining
Resins	Petrochemicals
Plastics	

Carbon	Coatings
Activated carbon regeneration	Paint
Graphite refractories	Ink
Electrodes	Dyes
Carbon black	Solvents

continued

TABLE 1.4 (CONTINUED)
Industries Requiring Thermal Oxidation Systems

Metals	Pulp and Paper
Furnace off-gas	Delac
Coking	TRS gas control
Cupola off-gas	Tissue drying
Pelletizing	Paper sludge disposal

Textiles	Minerals
Finishing	Pigments
Setting	Coal
Fabric manufacture	Kaolin
Treating	Ore roasting

Rubber	Sulfur
Tires	Acid gas regeneration
Moldings	Claus tail gas oxidation
Vulcanizing	
Butadiene off-gas	

Food	Pharmaceuticals
Corn fructose	Pill coating
Detergents	Vent gas
Perfume	

Miscellaneous

Spray propellants
Soil vapor treatment
Stripper off-gases
Engine exhaust dynamometers
Landfill gas combustion
Asphalt blow stills
Natural gas sweetening

2 Environmental Regulations

CONTENTS

Combustion was first discovered by the caveman. The smoke generated by this early combustion technology was a result of incomplete oxidation generating submicron particulate. In contrast, modern thermal oxidation systems represent controlled combustion.

Thermal oxidation systems owe their existence to environmental regulations. The regulations that have spurred the proliferation of thermal oxidizers are the 1990 Clean Air Act Amendments. The Clean Air Act (CAA) was originally passed by the U.S. Congress in 1955. The original act was amended in 1963 and 1965. The Air Quality act of 1967 provided the basic framework of the current statute. It established a basis for specifying acceptable levels of air pollution, required the federal government to specify air quality criteria, and required states to adopt air quality standards to meet these criteria. States were directed to designate air quality regions and set standards on a region-by-region basis.

The CAA was amended again in 1970. The amendments required the federal government to establish National Ambient Air Quality Standards (NAAQS) for six air pollutants. These criteria pollutants were sulfur oxides, particulate matter, ozone, carbon monoxide, nitrogen dioxide, and hydrocarbons. Primary NAAQS were intended to protect public health with an adequate margin for safety. Secondary NAAQS were intended to address harm to the environment. The 1970 amendments also established the New Source Performance Standards (NSPS) program that regulates emissions from new stationary sources. The 1970 amendments also gave the EPA authority to regulate hazardous air pollutants (NESHAP) not covered by NAAQS. Under the NESHAPs program, only seven chemicals were regulated: asbestos, benzene, beryllium, arsenic, mercury, radionuclides, and vinyl chloride. The 1977 amendments added two new programs: nonattainment and Prevention of Significant Deterioration (PSD). The nonattainment provisions apply to areas where ambient air levels exceed the safe levels established by the NAAQS. The PSD applies to areas that are in attainment and is intended to ensure that they remain in attainment.

Until 1990, the EPA attempted to reduce air emissions using a chemical-by-chemical approach. This approach was ineffective, leading to the 1990 amendments. In these amendments, Congress designated a list of 189 chemicals and elements as hazardous air pollutants (HAPs) whose emission to the atmosphere must be regulated.

2.1 FEDERAL LAW — STATE IMPLEMENTATION

The Clean Air Act is a federal law, but each individual state is responsible for implementing its provisions. The U.S. Environmental Protection Agency (EPA) translates the law enacted by Congress into specific procedures and emission limits that must be followed by every entity in the U.S. This ensures that all Americans have the same basic health and environmental protection. The law allows individual states to set more stringent standards, but no state can have standards less stringent than federal limits.

The law also recognizes that it makes sense for states to take the lead in implementing the Clean Air Act. The sources, characteristics, and potential effects of toxic air pollutants vary among individual states. In response to federal standards, each state must develop a plan to implement the federal standards. These State Implementation Plans (SIP) establish the control strategies, emission limitations, and timetables for compliance. The states must involve the public, through hearings, in the development of these plans.

2.2 1990 CLEAN AIR ACT TITLES

The Clean Air Act as amended in 1990 included a wide range of regulatory compliance issues. Each issue was differentiated under separate "Titles." These titles addressed ambient air levels, mobile (primarily automobile) emission sources, industrial emissions, acid rain, and enforcement actions among others. The original titles are listed in Table 2.1 which provides a brief synopsis of each.

2.2.1 TITLE I — NONATTAINMENT

Title I imposes emission limits on stationary sources (in contrast to the automobile). It has the greatest impact on what the public perceives to be air pollution. This title consists of two subprograms: National Ambient Air Quality Standards (NAAQS) and New Source Performance Standards (NSPS).

National Ambient Air Quality Standards (NAAQS)

Under the NAAQS program, the EPA has established what it considers to be safe levels of "criteria" pollutants in the ambient air. These criteria pollutants are ozone, particulate matter, carbon monoxide, sulfur oxides (SOx), nitrogen dioxide (NO_2), and lead. The trigger levels for these species were revised by the EPA on July 18, 1997 and are codified in 40 CFR Part 50 of the Code of Federal Regulations; they are shown in Table 2.2. Primary standards are intended to protect public health while secondary standards are intended to protect public welfare, including economic

---

TABLE 2.1
Clean Air Act Titles

Title I — Nonattainment
- Defines strategy for attainment of National Ambient Air Quality Standards (NAAQS)

Title II — Mobile Sources
- Intended to reduce tailpipe emissions from mobile sources

Title III — Air Toxics (NESHAP)
- Intended to reduce emissions of hazardous air pollutants from industrial sources

Title IV — Acid Rain
- Intended to reduce sulfur dioxide emissions primarily from power plants

Title V — Permits
- Defines permit requirements for each emission source

Title VI — Stratospheric Ozone Protection
- Establishes guidelines and schedules for phase-out of the use of chlorofluorocarbons

Title VII — Enforcement

Title VIII — Miscellaneous

Title IX — Clean Air Research

Title X — Disadvantaged Business Concerns

Title XI — Employment Transition Assistance

---

interests, vegetation, and visibility. The particulate matter limit applies to solids (or aerosols) in the air with aerodynamic diameters of both less than 2.5 microns (μm) and less than 10 microns. Also note that compliance with these standards is measured over varying averaging periods.

The EPA has divided the U.S. into regions and has classified these regions as either attainment, nonattainment, and unclassified. Unclassified regions are considered attainment until otherwise designated. If a region exceeds the specified ambient air levels for any one of the criteria pollutants, it is classified as nonattainment for that species. A region can be attainment for one species and nonattainment for another. For example, most regions are attainment for carbon monoxide but many are nonattainment for ozone. For ozone, carbon monoxide, and particulate, the EPA has further divided nonattainment areas into the subclassifications of marginal, moderate, serious, and extreme. Regulatory requirements in nonattainment areas become more stringent the greater the deviation from the prescribed ambient air attainment levels.

The 1990 Clean Air Act Amendments mandated that the EPA establish a program in such a manner that the criteria pollutant concentrations in nonattainment areas are reduced so that these areas eventually achieve attainment status. While the federal government sets the goals, individual states can decide how to achieve these goals within their jurisdiction. The states define their strategy for reaching attainment in nonattainment areas in their SIPs. The SIP defines the controls that sources must impose to reduce emissions in an effort to meet NAAQS guidelines.

TABLE 2.2
National Ambient Air Quality Standards

Criteria Pollutant	Type of Standard	NAAQS Limit µg/m³	ppm	Measurement Period
Ozone	Primary, secondary	—	0.08	3-yr av of 4th highest daily max of 8-hr av
	Primary, secondary (Only in ozone nonattainment zones)	235	0.12	Max hourly av; one allowable exceedance per year
Carbon monoxide	Primary	10,000	9	8-hr av — one allowable exceedance per year
	Primary	40,000	35	1-hr av conc; one allowable exceedance per year
Particulate matter (PM-2.5)	Primary and secondary	65	–	3-yr av of 98th percentile 24-hr avs
	Primary and secondary	15	–	3-yr av of annual mean conc
Particulate matter (PM-10)	Primary and secondary	150	–	3-yr av of 98th percentile 24-hr avs
	Primary and secondary	150	–	Max 24-hr conc; one exceedance allowed per year
	Primary and secondary	15	–	3-yr av of annual mean conc
Sulfur dioxide (SOx) (as SO_2)	Primary	80	0.03	Annual mean
	Primary	365	0.14	Max-24 hr conc; one exceedance allowed per year
	Secondary	1,300	0.5	Max 3-hr conc; one exceedance allowed per year
Nitrogen dioxide	Primary and secondary	100	0.053	Annual mean (as NO_2)
Lead	Primary and secondary	1.5	–	Max mean over a calendar quarter

Note: In May 1999, the U.S. Court of Appeals for the District of Columbia voided the ozone and PM-2.5 standards. The EPA has appealed that court's decision.

Source: *The Air Pollution Consultant, Quick Reference Guide*, Vol. 8, Issue 7, December 1998 (Elsevier Science). With permission.

Title I of the Clean Air Act amendments of 1990 has a direct impact on the need for a thermal oxidizer. While VOCs are not among the six criteria pollutants on which the ambient air levels are based, VOCs in combination with nitrogen oxides (NOx) in the atmosphere can combine to form ozone. More areas of the U.S. are classified as nonattainment due to ambient ozone concentrations than any of the other six criteria pollutants. This is the primary reason for the emphasis on reduction of VOC emissions to the atmosphere.

2.2.2 NEW SOURCE PERFORMANCE STANDARDS (NSPS)

The NSPS program regulates emissions from new or modified stationary sources of air pollutant emissions. These technology-based standards apply regardless of the attainment status of the area where the source is located. They reflect the degree of emissions reduction achievable through application of the best system of emission reduction that the EPA determines has been adequately demonstrated for that category of sources. NSPS have been promulgated for more than 70 source categories (industries), "taking into account the cost of achieving such reductions and any non-air quality health and environmental impact and energy requirements the Administrator has deemed adequately demonstrated" (CAA Section 111 (a)(1)). This concept is known as "best demonstrated technology" (BDT). Within each source category, specific pollutants are identified along with emission standards.

2.2.3 TITLE II — MOBILE SOURCES

Automobiles account for about 50% of the VOC and NOx and 90% of the carbon monoxide emitted to the atmosphere in urban areas. Congress mandated that the EPA revise tailpipe emission standards for both cars and trucks. Tier I standards, phased in beginning with the 1994 model year, limited tailpipe emissions of hydrocarbons, NOx, and carbon monoxide. The EPA was to consider the need for more stringent standards (Tier II) by the end of 1999. The chemical and physical properties of gasoline were also affected. This included the use of unleaded, reformulated, and oxygenated gasolines. The sale of gasoline with a Reid vapor pressure greater than 9 psi is also banned during the high ozone season. This title also includes provisions to limit the sulfur content of diesel fuel. The EPA is also studying the need to control emission of HAPs from mobile sources and set HAP emission standards.

2.2.4 TITLE III — AIR TOXICS

No title of the CAA has had such a direct impact on the proliferation of thermal oxidation systems as Title III. This title replaced the original NESHAP program. This title requires standards to be established for categories and subcategories of sources that emit certain chemical species to the atmosphere. It differs from the original program in that sources are regulated, rather than the pollutants themselves. The intent was to develop a single set of standards, which would cover all listed hazardous air pollutants.

Congress listed 189 specific chemical compounds in the 1990 amendments whose emissions were to be reduced. Subsequently, two compounds were removed from the list (caprolactam and methyl acetate). This list is shown in Table 2.3. These compounds are designated as HAPs. The intent of the 1990 CAA amendments was to reduce emission of these HAPs by 90%. The EPA was also directed by Congress to review this list periodically and add compounds that "present or may present, through inhalation or other routes of exposure, a threat of adverse human health effects (including, but not limited to, substances which are known to be, or may be reasonably anticipated to be, carcinogenic, mutagenic, teratogenic,

TABLE 2.3
Hazardous Air Pollutants

CAS Number	Name	CAS Number	Name
75070	Acetaldehyde	156627	Calcium cyanamide
60355	Acetamide	133062	Captan
75058	Acetonitrile	63252	Carbaryl
98862	Acetophenone	75150	Carbon disulfide
53963	2-Acetylaminofluorene	56235	Carbon tetrachloride
107028	Acrolein	463581	Carbonyl sulfide
79061	Acrylamide	120809	Catechol
79107	Acrylic acid	133904	Chloramben
107131	Acrylonitrile	57749	Chlordane
107051	Allyl chloride	7782505	Chlorine
92671	4-Aminobiphenyl	79118	Chloroacetic acid
62533	Aniline	532274	2-Chloroacetophenone
90040	o-Anisidine	108907	Chlorobenzene
1332214	Asbestos	510156	Chlorobenzilate
71432	Benzene	67663	Chloroform
92875	Benzidine	107302	Chloromethyl methyl ether
98077	Benzotrichloride	126998	Chloroprene
100447	Benzyl chloride	1319773	Cresols/cresylic acid
92524	Biphenyl	95487	o-Cresol
117817	Bis(2-ethylhexyl)phthalate	108394	m-Cresol
542881	Bis(chloromethyl)ether	106445	p-Cresol
75252	Bromoform	98828	Cumene
106990	1,3 Butadiene	94757	2,4 D, salts and esters
3547044	DDE	534521	4,6-Dinitro-o-cresol
334883	Diazomethane	51285	2,4-Dinitrophenol
132649	Dibenzofurans	121142	2,4-Dinitrotoluene
96128	1,2 Dibromo-3-chloropropane	123911	1,4-Dioxane
84742	Dibutylphthalate	122667	1,2-Diphenylhydrazine
106467	1,4 Dichlorobenzene	106898	Epichlorohydrin
91941	3,3-Dichlorobenzidene	106887	1,2 - Epoxybutane
111444	Dichloroethyl ether	140885	Ethyl acrylate
542756	1,3-Dichloropropene	100414	Ethyl benzene
62737	Dichlorvos	51796	Ethyl carbamate
111422	Diethanolamine	75003	Ethyl chloride
121697	N,N-Diethyl aniline	106934	Ethylene dibromide
64675	Diethyl sulfate	107062	Ethylene dichloride
119904	3,3-Dimethoxybenzidene	107211	Ethylene glycol
60117	Dimethyl amoniazobenzene	151564	Ethylene imine
119937	3,3-Dimethyl benzidene	75218	Ethylene oxide
79447	Dimethyl carbamoyl chloride	96457	Ethylene thiourea

TABLE 2.3 (CONTINUED)
Hazardous Air Pollutants

68122	Dimethyl formamide	75343	Ethylidene dichloride
7147	1,1-Dimethyl hydrazine	50000	Formaldehyde
131113	Dimethyl phthalate	76448	Heptachlor
77781	Dimethyl sulfate	118741	Hexachlorobenzene
87683	Hexachlorobutadiene	60344	Methyl hydrazine
77474	Hexachlorocyclopentadiene	74884	Methyl iodidie
67721	Hexachloroethane	108101	Methyl isobutyl ketone
822060	Hexamethylene-1,6-diisocyanate	624839	Methyl isocyanate
680319	Hexamethylphosphoramide	80626	Methyl methacrylate
110543	Hexane	1634044	Methyl *tert*-butyl ether
302012	Hydrazine	101144	4,4-Methylene bis(2-chloroaniline)
7647010	Hydrochloric acid	75092	Methylene chloride
7664393	Hydrogen fluoride	101688	Methylene diphenyl diisocyanate
7783064	Hydrogen sulfide	101779	4,4-Methylenedianiline
123319	Hydroquinone	91203	Napthalene
78591	Isophorone	98953	Nitrobenzene
58899	Lindane	92933	4-Nitrobiphenyl
108316	Maleic anhydride	100027	4-Nitropropane
67561	Methanol	684935	*N*-Nitroso-*N*-methylurea
72435	Methoxychlor	62759	*N*-Nitrosodimethylamine
74839	Methyl bromide	59892	N-Nitrosomorpholine
74873	Methyl chloride	56382	Parathion
71556	Methyl chloroform	82688	Pentachloronitrobenzene
78933	Methyl ethyl ketone	87865	Pentachlorophenol
108952	Phenol	127184	Tetrachloroethylene
106503	*p*-Phenylenediamine	7550450	Titanium tetrachloride
75445	Phosgene	108883	Toluene
7803512	Phosphine	95807	2,4-Toluene diamine
7723140	Phosphorous	584849	2,4-Toluene diisocyanate
85449	Phthalic anhydride	95534	*o*-Toluidine
1336363	Polychlorinated biphenyls	8001352	Toxaphene
1120714	1,3-Propane sulfone	120821	1,2,4-Trichlorobenzene
57578	beta-Propiolactone	79005	1,1,2-Trichloroethane
123386	Propionaldehyde	79016	Trichloroethylene
114261	Propoxur	95954	2,4,5-Trichlorophenol
78875	Propylene dichloride	88062	2,4,6-Trichlorophenol
75569	Propylene oxide	121448	Triethylamine
75558	1,2-Propylenimine	1582098	Trifluralin

continued

TABLE 2.3 (CONTINUED)
Hazardous Air Pollutants

81225	Quinoline	540841	2,2,4-Trimethylpentane
106514	Quinone	108054	Vinyl acetate
100425	Styrene	593602	Vinyl bromide
96093	Styrene oxide	75014	Vinyl chloride
1746016	2,3,7,8-Tetrachlorodibenzo-*p*-dioxin	75354	Vinylidene chloride
79345	1,1,2,2-Tetrachloroethane	1330207	Xylenes (isomers and mixtures)
95476	*o*-Xylenes	108383	*m*-Xylene
106423	*p*-Xylene		

Antimony compounds	Manganese compounds
Mercury compounds	Fine mineral fibers
Nickel compounds	Polycyclic organic matter
Radionuclides	Selenium compounds
Arsenic compounds	Beryllium compounds
Cadmium compounds	Chromium compounds
Cobalt compounds	Coke oven emissions
Cyanide compounds	Glycol ethers
Lead compounds	Mercury compounds
Fine mineral fibers	Nickel compounds
Polycyclic organic matter	Radionuclides
Selenium compounds	

Source: Section 112(b)(1) of the 1990 Clean Air Act Amendments.

neurotoxic, which cause reproductive dysfunction, or which are acutely or chronically toxic."

Under this title, the EPA developed a list of 166 major source categories and 8 area source categories. A major source is defined as any single source or group of stationary sources located within a contiguous area that has the potential to emit 10 tons per year of any single HAP or 25 tons per year of a combination of HAPs. An area source is defined as any stationary source of HAPs that is not a major source. The source categories established by the EPA and the schedule for finalizing MACT standards for that category are shown in Table 2.4. Note, for a variety of reasons, standards are not always promulgated when scheduled.

To control the emission of HAPs from these source categories, the EPA has established (and continues to establish) Maximum Achievable Control Technology (MACT) standards. For new sources, MACT represents the best controlled similar source. For existing sources, MACT is defined in two ways, depending on the prevalence of that type of source. For categories with 30 or more existing sources, MACT is defined as controls that are no less stringent than the average emissions levels achieved by the best performing 12% of sources in that particular source category. For categories with less than 30 sources, MACT represents the average

TABLE 2.4
MACT Categories and Schedule for Finalizing Standard

Category	Proposed or Actual Date of Standard Promulgation
Agricultural chemical production	
Butadiene-furfural cotrimer production	4/24/94
Captafol production	4/22/94
4-Chloro-2-methylphenoxyacetic acid	11/15/97
Chloroneb production	11/15/97
Chlorothalonil production	4/22/94
2,4-D salt and ester production	11/15/97
Dacthal™ production	4/22/94
4,6 Dinitro-o-cresol production	11/15/97
Sodium pentachlorophenate production	11/15/97
Tordon™ acid production	5/22/94
Ferrous metal processing	
Coke by-products plants	11/15/00
Coke ovens — charging, top side, and door leaks	10/27/93
Coke ovens — pushing, quenching, and battery stacks	11/15/00
Ferroalloy production	11/15/97
Integrated iron and steel manufacturing	11/15/00
Iron foundries	11/15/00
Steel foundries	11/15/00
Steel pickling — HCl process	11/15/97
Fiber production process	
Acrylic fiber/modacrylic fiber production	11/15/97
Rayon production	11/15/00
Spandex production	11/15/00
Food and agriculture processes	
Baker's yeast manufacturing	11/15/00
Cellulose food casing manufacturing	11/15/00
Vegetable oil production	11/15/00
Fuel combustion	
Engine test facilities	11/15/00
Industrial boilers	11/15/00
Institutional/commercial boilers	11/15/00
Process heaters	11/15/00
Stationary internal combustion engines	11/15/00
Liquid distribution	
Gasoline distribution	12/14/94
Marine vessel loading	9/19/95
Organic liquid distribution	11/15/00
Mineral product processing	
Alumina processing	11/15/00
Asphalt/coal tar applications — metal pipes	11/15/00

continued

TABLE 2.4 (CONTINUED)
MACT Categories and Schedule for Finalizing Standard

Category	Proposed or Actual Date of Standard Promulgation
Asphalt concrete manufacturing	11/15/00
Asphalt processing	11/15/00
Asphalt roof manufacturing	11/15/00
Chromium refractory production	11/15/00
Clay product manufacturing	11/15/00
Lime manufacturing	11/15/00
Mineral wool production	11/15/97
Portland cement manufacturing	11/15/97
Taconite iron ore processing	11/15/00
Wood fiberglass manufacturing	11/15/97
Nonferrous metal processing	
Primary aluminum production	10/7/97
Secondary aluminum production	11/15/97
Primary copper smelting	11/15/97
Primary lead smelting	11/15/97
Secondary lead smelting	6/23/95
Primary magnesium refining	11/15/00
Petroleum and natural gas production and refining	
Oil and natural gas production	11/15/97
Natural gas transmission and storage	11/15/97
Petroleum refineries — catalytic cracking, reforming, and sulfur plant units	11/15/97
Petroleum refineries — other sources not listed	8/18/95
Pharmaceutical production	9/21/98
Polymer and resin production	
Acetal resin	11/15/97
Acrylonitrile–butadiene–styrene (ABS) resin	9/12/96
Alkyd resin	11/15/00
Amino resin	11/15/97
Boat manufacturing	11/15/00
Butyl rubber	9/5/96
Carboxymethylcellulose	11/15/00
Cellulose ether	11/15/00
Epichlorohydrin elastomer	9/5/96
Epoxy resin	3/8/95
Ethylene–propylene rubber	9/5/96
Flexible polyurethane foam	10/7/98
Hypalon	9/5/96
Maleic anhydride copolymer	11/15/00
Methylcellulose	11/15/00
Methyl methacrylate–acrylonitrile–butadiene–styrene resin	9/12/96

TABLE 2.4 (CONTINUED)
MACT Categories and Schedule for Finalizing Standard

Category	Proposed or Actual Date of Standard Promulgation
Methyl methacrylate–acrylonitrile–butadiene–styrene terpolymer	9/12/96
Neoprene	9/5/96
Nitrile butadiene rubber	9/5/96
Nitrile resins	9/12/96
Nonnylon polyamide	3/8/95
Phenolic resin	11/15/97
Polybutadiene rubber	9/5/96
Polycarbonate	4/22/94
Polyester resin	11/15/00
Polyether polyol	11/15/97
Polyethylene terephthalate resin	9/12/96
Polymerized vinylidene chloride	11/15/00
Polymethyl methacrylate resin	11/15/00
Polystyrene resin	9/12/96
Polysulfide rubber	9/5/96
Polyvinyl acetate emulsion	11/15/00
Polyvinyl alcohol	11/15/00
Polyvinyl butyral	11/15/00
Polyvinyl chloride and copolymer	11/15/00
Reinforced plastic composites	11/15/00
Styrene–acrylonitrile resin	9/12/96
Styrene–butadiene rubber and latex	9/5/96
Production of inorganic chemicals	
Ammonium sulfate — caprolactam by-products plants	11/15/00
Antimony oxide	11/15/00
Carbon black	11/15/00
Chlorine	11/15/00
Cyanide chemical	11/15/00
Fume silica	11/15/00
Hydrochloric acid	11/15/00
Hydrogen fluoride	11/15/00
Phosphate fertilizer	11/15/97
Phosphoric acid	11/15/97
Uranium hexafluoride	11/15/00
Production of organic chemicals	
Ethylene processes	11/15/00
Quarternary ammonium compound	11/15/00
Synthetic organic chemical manufacturing	4/22/94

continued

TABLE 2.4 (CONTINUED)
MACT Categories and Schedule for Finalizing Standard

Category	Proposed or Actual Date of Standard Promulgation
Surface coating processes	
Aerospace industries	9/1/95
Auto and light duty truck (surface coating)	11/15/00
Flat wood paneling (surface coating)	11/15/00
Large appliance (surface coating)	11/15/00
Magnetic tapes (surface coating)	12/15/94
Manufacture of paints, coatings, and adhesives	11/15/00
Metal can (surface coating)	11/15/00
Metal coil (surface coating)	11/15/00
Metal furniture (surface coating)	11/15/00
Miscellaneous metal parts and products (surface coating)	11/15/00
Paper and other webs (surface coating)	11/15/00
Plastic parts and products (surface coating)	11/15/00
Printing, coating, and dyeing of fabrics	11/15/00
Printing/publishing (surface coating)	5/30/96
Shipbuilding and ship repair (surface coating)	12/15/95
Wood furniture (surface coating)	12/7/95
Waste treatment and disposal	
Hazardous waste incineration	4/19/96
Off-site waste and recovery operations	7/1/96
Publicly owned treatment works emissions	11/15/95
Sewage sludge incineration	11/15/00
Site remediation	11/15/00
Manufacturing processes	
Aerosol can-filling facilities	11/15/00
Benzyltrimethylammonium chloride production	11/15/00
Carbonyl sulfide production	11/15/00
Chelating agent production	11/15/00
Chlorinated parrafin production	4/22/94
Chromic acid anodizing	1/25/95
Commercial sterilization facilities	12/6/94
Decorative chromium electroplating	1/25/95
Dry cleaning (perchloroethylene)	9/22/93
Dry cleaning (petroleum solvents)	11/15/00
Ethylidene norbornene production	4/22/94
Explosives production	11/15/00
Flexible polyurethane foam fabrication operations	11/15/00
Friction product manufacturing	11/15/00
Halogenated solvents cleaners	12/2/94
Hard chromium electroplating	1/25/95
Hydrazone production	11/15/00

TABLE 2.4 (CONTINUED)
MACT Categories and Schedule for Finalizing Standard

Category	Proposed or Actual Date of Standard Promulgation
Industrial process cooling towers	9/8/94
Leather tanning and finishing operations	11/15/00
OBPA/1,3-diisocyanate production	4/22/94
Paint stripper users	11/15/00
Photographic chemical production	11/15/00
Phthalate plasticizer production	4/22/94
Plywood/particle board manufacturing	11/15/00
Pulp and paper production	
Kraft, soda, sulfite, semichemical, mechanical pulping; and nonwood fiber pulping	4/15/98
Chemical recovery combustion sources	11/15/00
Rocket engine test firing	11/15/00
Rubber chemical manufacturing	11/15/00
Semiconductor manufacturing	11/15/00
Symmetrical tetrachloropyridine production	4/22/94
Tire production	11/15/00

Source: *The Air Pollution Consultant, Quick Reference Guide*, Vol. 8, Issue 7, December 1998 (Elsevier Science).

emission limit achieved by the best performing five sources. In a limited number of cases, the EPA has produced Control Techniques Guidance (CTG) documents that identify control technologies that can be applied to specific source categories to meet the MACT limits.

For area sources, the EPA established Generally Available Control Technology (GACT). GACT is less stringent than MACT. The EPA was directed by Congress to review and revise, if applicable, MACT standards every 8 years. In general, existing sources must comply with MACT standards within 3 years of promulgation. New sources must comply at either the later of the start-up date or the standard's effective date.

2.2.5 TITLE IV — ACID RAIN

The title is primarily aimed at reducing emissions of NOx and sulfur dioxide from coal-burning electric utilities. These chemical species are believed to be precursors to acid rain. The SO_2 and NOx released are converted to sulfuric acid and nitric acid mists in the atmosphere. When winds blow these acid chemicals into areas where the weather is wet, they become part of the rain, snow, or fog. Lakes and streams are normally slightly acidic, but acid rain can make them very acidic, with consequent damage to plant and animal life.

While certain provisions of this title of the CAA were implemented in 1995, most utility will be required to reduce SO_2 emissions to 1.2 pounds per MM Btu of fuel consumption by the year 2000. The NOx and SO_2 limits can be achieved in different ways under the EPA program. While NOx emissions are to be met using traditional combustion and postcombustion control techniques, the SO_2 limits can be met through a program of market allowances. Under this program, each installation is allocated emission allowances that can be bought, sold, or shared with other utilities. Units are not allowed to emit more SO_2 than the allowances they hold. Utilities that have low emission rates can sell their excess allowances to other utilities or use them to expand capacity. Beginning January 1, 2000, utilities will be assigned an aggregate of 8.9 million tons of SO_2 emissions allowances.

To verify that emission limits are being met, utilities are required to install continuous emission monitoring systems (CEMS) to determine total SO_2 and NOx emissions. Extensive recordkeeping is also required to document these emissions.

2.2.6   TITLE V — OPERATING PERMITS

The 1990 CAA requires that most air emission sources obtain an operating permit. The intent of this program was not to create additional requirements for a facility, but to compile applicable requirements from all CAA programs in one document. While this is again a federal EPA-mandated program, it is administered by each state under its State Implementation Plan. Each state was required to submit a detailed plan to the EPA for approval of its SIP before implementing the Title V requirements. This submission had to show that the state program included specific regulations, statutes, and enforcement authority. However, each state was free to create its own version, as long as it complied with EPA mandates. As with other provisions of the CAA, state programs could be more stringent than federal requirements. Thus, many states included additional provisions in their operating permits program.

As instructed in the SIP, each source must file a permit application that includes documentation that all CAA requirements are being met. This was one of the more contentious provisions of the CAA, delaying final rulemaking until 1997.

2.2.7   TITLE VII — ENFORCEMENT

Another of the more contentious provisions of the Clean Air Act Amendments of 1990 deals with enforcement. To demonstrate that facilities were meeting emission limits, the EPA proposed an enhanced monitoring program. This program would have required sources to install and operate CEMS to ensure compliance. Based on its cost and complexity, industry strenuously objected to its implementation. As a result, the EPA withdrew this program in favor of a less burdensome program called Compliance Assurance Monitoring (CAM). Under this program, a source must demonstrate compliance with CAA emission limits by either measuring emission rates or by controlling and recording operating parameters of a control device.

While not all of the CAA titles have been discussed here, the remainder do not have a significant impact on the need, design, or operation of a thermal oxidation system.

3 VOC Destruction Efficiency

CONTENTS

Precisely what is a volatile organic compound (VOC)? According to the Code of Federal Regulations (40 CFR 51.100), a VOC is a compound of carbon, excluding carbon monoxide, which participates in atmospheric photochemical reactions. Certain compounds have been specifically excluded in the federal definition of a VOC. These include methane, ethane, acetone, carbon dioxide, carbonic acid, metallic carbides or carbonates, and ammonium carbonate methylene chloride, trichloroethane, and many chlorofluorocarbons.

The proliferation of thermal oxidizers for treatment of waste streams containing volatile organic compounds is based on their ability to destroy rather than just capture (requiring further treatment) the VOCs. However, thermal oxidation is a chemical reaction. Very few chemical reactions go to 100% completion. Therefore, the performance of a thermal oxidizer is usually judged in terms of its approach to 100% oxidation of the VOCs. This approach to complete oxidation is defined as the destruction efficiency (DE) or, alternately, destruction removal efficiency (DRE). It is the mass ratio of the VOCs fed to the thermal oxidizer minus the unreacted VOCs in the stack emissions divided by the VOC feed rate. In equation form:

$$\text{VOC destruction efficiency} = \frac{(\text{VOCs fed to thermal oxidizer} - \text{VOCs in stack gas})}{\text{VOCs fed to thermal oxidizer}} \times 100 \tag{3.1}$$

For example, if a waste stream contains 100 lb/hr of a combination of VOCs, and analysis of the thermal oxidizer stack gas reveals that 1 lb/hr remains unoxidized, the destruction efficiency is 99%. From Equation 3.1:

$$\text{VOC destruction efficiency} = \frac{(100 \text{ lb/hr} - 1 \text{ lb/hr})}{100 \text{ lb/hr}} \times 100 = 99\%$$

Destruction efficiency should be distinguished from capture and destruction. If a process releases 100 lb/hr of a particular (or combination) of VOCs, and the system used to collect these VOCs is only able to capture 98%, then the collection system has a 98% capture efficiency. This must be factored together with the ultimate VOC destruction efficiency of the thermal oxidizer to obtain the overall VOC capture and destruction efficiency. If 99% VOC destruction is achieved in the thermal oxidizer, then the overall capture and removal efficiency is $0.98 \times 0.99 \times 100 = 97\%$.

3.1 OPERATING PARAMETERS

Thermal oxidation systems can achieve very high levels of VOC destruction when properly designed and operated. The parameters that define these optimum conditions have been historically described as the "three Ts of destruction": time, temperature, and turbulence. A fourth must also be included, excess oxygen. VOC destruction efficiencies greater than 99.99% can be achieved if these four parameters are set in the proper range.

3.1.1 TEMPERATURE

No parameter has a greater impact on VOC destruction than the operating temperature of the thermal oxidizer. Generally, thermal oxidizers operate in a temperature range of 1400 to 2200°F. The exception is the treatment of a waste stream containing total reduced sulfur (TRS) compounds. These compounds can be effectively destroyed at temperatures as low as 1200°F. Examples of TRS compounds are hydrogen sulfide, methyl mercaptan, and dimethyl sulfide. At a given operating temperature, destruction efficiency will vary with the specific compound treated. That is, higher destruction efficiency will be achieved for some compounds in comparison to others at the same temperature.

One generally accepted method of estimating the temperature required for destruction of an organic compound is its autoignition temperature (AIT). The AIT temperature of a compound is the temperature above which a flammable mixture is capable of extracting enough energy from the environment to self-ignite. Compounds with higher AITs are usually more difficult to destroy. AITs of common VOCs are shown in Table 3.1.

3.1.2 RESIDENCE TIME

Residence time is another of the three Ts of destruction. However, it does not have the same impact as temperature on VOC destruction efficiency. Nonetheless, sufficient time must be allowed for the chemical kinetic reactions to occur. Generally, thermal oxidizer residence times range from 0.5 to 2.0 s. Lower residence times correspond to lower destruction efficiencies and vice versa. A 1.0-s residence time is generally applied when a destruction efficiency of 99.99% or higher is required.

The thermal oxidizer operating temperature can be lowered if the residence time is increased. However, since temperature is the predominant factor affecting VOC destruction, this reduction is usually not more than 50 to 100°F. In some situations, the operating temperature and residence time are set by environmental regulations. One example is the Toxic Substances Control Act (TSCA) which regulates wastes containing polychlorinated biphenyls (PCBs). It requires a temperature of 2192°F (1200°C) with a 2-s gas residence time and 3% oxygen in the combustion products.

TABLE 3.1
Autoignition (AIT) Temperatures of Common Organic Compounds

Compound	Autoignition Temperature (°F)	Compound	Autoignition Temperature (°F)
Acetone	869	Hydrogen sulfide	500
Ammonia	1204	Kerosene	490
Benzene	1097	Maleic anhydride	890
Butadiene	840	Methane	999
Butanol	693	Methanol	878
Carbon disulfide	257	Methyl ethyl ketone	960
Carbon monoxide	1128	Methylene chloride	1224
Chlorobenzene	1245	Mineral spirits	475
Dichloromethane	1185	Petroleum naphtha	475
Dimethyl sulfide	403	Nitrobenzene	924
Ethane	950	Phthalic anhydride	1084
Ethyl acetate	907	Propane	874
Ethanol	799	Propylene	940
Ethylbenzene	870	Styrene	915
Ethyl chloride	965	Trichloroethane	932
Ethylene dichloride	775	Toluene	997
Ethylene glycol	775	Turpentine	488
Hydrogen	1076	Vinyl acetate	800
Hydrogen cyanide	1000	Xylene	924

3.1.3 TURBULENCE

Oxygen and VOC molecules must be thoroughly mixed at the prescribed temperature for the chemical oxidation reactions to approach completion. This is accomplished by ensuring a high degree of turbulence within the thermal oxidizer. The gas Reynolds number (Re) generally defines turbulence. It is calculated for a thermal oxidizer as follows:

$$Re = \frac{(\text{Oxidizer internal diameter}) \times (\text{gas velocity}) \times (\text{gas density})}{(\text{Gas viscosity})}$$

To ensure complete turbulence, the Re should be greater than 10,000.

This concept can be simplified by recognizing that some of the parameters in the Re equation are interrelated. For example, velocity is dependent upon the thermal oxidizer inner diameter. Velocity, density, and viscosity are dependent on temperature. Furthermore, the compositions of the products of combustion generally fall within a fairly narrow range. Thus, the density and viscosity also vary within a very narrow range for a given temperature. Not intuitively obvious is the fact that higher velocities are required at higher temperatures. This occurs because the gas density decreases and viscosity increases at higher temperatures. As a rule of thumb, gas velocities should be maintained above 20 ft/s.

3.1.4 OXYGEN CONCENTRATION

The concentration of oxygen molecules is another important component of thermal oxidation reactions. Oxygen is generally supplied by the addition of combustion air or, in the case of a VOC-contaminated air stream, may be present as part of the waste stream itself. To ensure that VOC molecules come in contact with oxygen molecules, excess oxygen is supplied to the system. Typically, this excess oxygen is established by maintaining an oxygen concentration in the products of combustion of at least 3.0%.

3.2 DESTRUCTION EFFICIENCY

VOC destruction rates are difficult to quantify from a purely theoretical standpoint. A statistical model has been proposed from laboratory studies.[2] This model relates design and operating parameters with VOC chemical and physical properties. However, this model was developed under plug flow conditions that do not exist in real systems. It also only applies to destruction efficiencies of 99% or greater.

Waste gas stream characteristics can vary over a wide range. Selection of thermal oxidizer operating parameters to achieve optimum VOC destruction is best left to companies that have accumulated years of operating data at a variety of conditions. However, with this caveat, Table 3.2 provides guidelines for VOC destruction efficiency as a function of temperature and residence time. This table assumes that at least 3.0% oxygen concentration is present in the products of combustion and that sufficient turbulence, as defined above, is present.

3.3 EPA INCINERABILITY RANKING

If a thermal oxidizer is operated at conditions which destroy most of the VOCs present, why aren't all destroyed? Theory suggests that oxygen-starved pathways are responsible for incomplete oxidation. Even though the oxidizer seems to be operating under excess air conditions, there are localized zones within the thermal oxidizer which are oxygen deficient. It is this theory from which the EPA developed an incinerability ranking. Gas-phase thermal stability in oxygen-deficient atmospheres is considered representative of the relative thermal stability of organic compounds. Based on research conducted at the University of Dayton Research

TABLE 3.2
VOC Destruction Efficiency vs. Time
and Temperature

Destruction Efficiency (%)	Degrees (°F) Above AIT	Residence Time (s)
95	300	0.5
98	400	0.5
99	475	0.75
99.9	550	1.0
99.99	650	2.0

Institute, a ranking of the thermal stability of hundreds of organic compounds was developed. This ranking does not define the exact conditions required to achieve a given level of thermal destruction, but makes a comparison of the degree of difficulty of thermal destruction of organic compounds.

This Incinerability Ranking is shown in Appendix A.[3] Compounds are grouped by class. Those within a given class are similar in their resistance to thermal destruction. For instance, hydrogen cyanide and benzene are in Class I. Both are considered relatively difficult to destroy using thermal oxidation technology. Toluene is in Class II. Therefore, it is more amenable to thermal destruction than either benzene or hydrogen cyanide. Ethylene dibromide is in Class 5. Based on its incinerability ranking, it is considered easier to thermally destruct than toluene or hydrogen cyanide or benzene. Again, this ranking does not specify the conditions needed to achieve a given degree of destruction, only the relative difficulty in destroying an organic compound. If actual data are available for the destruction efficiency of a particular compound, then the incinerability ranking can be used to determine whether the destruction efficiency of another compound will be higher or lower at these same conditions.

The incinerability ranking was developed for use in designing hazardous waste incinerators. However, the data also apply to thermal oxidizers. While most of the compounds listed in Appendix A would not be encountered with thermal oxidation because they are solids or liquids or are relatively obscure compounds, there are some that are relevant. Examples are benzene, hydrogen cyanide, naphthalene, acetonitrile, acrylonitrile, toluene, aniline, pyridine, nitrobenzene, and many chlorinated organic compounds.

3.4 ENVIRONMENTAL REGULATIONS

As discussed in Chapter 2, Maximum Achievable Control Technology (MACT) standards have been and are continuing to be promulgated for 174 specific source categories for VOC emissions. Finalized MACT standards have been promulgated for the synthetic organic chemical manufacturing industry (SOCMI) and the pulp and paper industry. Both establish combustion devices as reference control technology (RCT) for certain wastes. In these cases, combustion devices are the technology standard against which competing technologies are measured.

For the SOCMI industry, the MACT standard is known as the HON (hazardous organic NESHAP) rule. For combustion devices, compliance requires 98% VOC destruction. The proposed pulp and paper industry MACT standard also specifies 98% VOC destruction. New Source Performance Standards (NSPS) have also been proposed for VOCs emanating from wastewater treatment operations. For VOC thermal oxidation, 95% destruction is required. This level is presumed to be achieved if the oxidizer is operated at 1400°F with a 0.5-s gas residence time. In the case of the pulp and paper MACT standard, 98% destruction efficiency is presumed if a thermal oxidizer is operated at specified conditions. These are a temperature of 1600°F and a gas residence time of 0.75 s.

MACT standards have also been proposed for commercial, off-site, waste treatment, storage, and disposal facilities. Control devices must reduce total organic carbon (TOC) or total hazardous air pollutant (HAP) emissions by 95%. If a thermal oxidizer is used as the control device, it must be operated at 1400°F with a gas residence time of 0.5 s.

Generally, thermal oxidizers are not operated at a temperature of less than 1400°F. Some state regulations require a minimum operating temperature of 1500°F, regardless of the VOC components of the waste stream. There is one exception: waste gases whose only objectionable components are sulfur compounds. Examples are hydrogen sulfide, methyl mercaptan, dimethyl sulfide, dimethyl disulfide, and carbonyl sulfide. None of these compounds have an AIT above 500°F. Therefore, high destruction efficiencies can be achieved with thermal oxidizer temperatures as low as 1200°F.

3.5 HALOGENATED COMPOUNDS

In the periodic table of the elements, elements are grouped in columns that have similar properties. One such group is called the halogens. The elements that comprise halogens are fluorine, chlorine, bromine, iodine, and astatine. They are mentioned here because many organic compounds contain one or more of these halogen elements, the most common being chlorine.

Thermal oxidation of compounds containing halogens requires special considerations. In general, these compounds are among the most difficult to destroy. Examples are chlorinated solvents (e.g., methylene chloride, chlorobenzene, dichloromethane, and trichloroethane). This difficulty in destruction is reflected in their high AITs.

There are additional considerations when destroying halogenated compounds. One is acid gas emissions. For example, with chlorinated compounds, the chlorine atoms in the original VOC are converted to a mixture of primarily hydrogen chloride gas (HCl), plus a small amount of chlorine gas (Cl_2).

Environmental regulations limit chlorine (HCl or Cl_2) emissions to the atmosphere. These compounds can be removed with an acid gas scrubber downstream of the thermal oxidizer. Hydrogen chloride can be removed simply by scrubbing with water, while chlorine gas requires a caustic agent such as caustic soda (sodium hydroxide). By operating the thermal oxidizer at a higher temperature, the proportion of chlorine gas generated vs. hydrogen chloride decreases. This relative proportion is determined by chemical equilibrium. Chemical equilibrium will be discussed in more detail in Chapter 4, along with acid dewpoint considerations.

Example 3.1: A batch reactor vent gas contains acetone, dimethylamine, and ethyl acetate together with nitrogen, oxygen, carbon dioxide, and water vapor. What residence time and temperature are required to achieve 99% destruction of all components?

Solution: The autoignition temperatures of these organic components are as follows:

Acetone	869°F
Dimethylamine	594°F
Ethyl acetate	907°F

The criteria for 99% destruction from Table 3.2 are an operating temperature 475°F above the autoignition temperature, with a gas residence time inside the combustion chamber of 0.75 s. Since ethyl acetate has the highest AIT, an operating temperature of 1500°F is selected (475 + 907 = 1482°F).

Example 3.2: In an air stream contaminated with benzene, 99% of the benzene present must be thermally destroyed. What thermal oxidizer operating conditions are required?

Solution: The autoignition temperature of benzene is 1097°F. Based on the criteria from Table 3.2, the operating temperature must be at least 1600°F (1097 + 475 = 1572°F). The combustion chamber must be sized to achieve a 0.75-s gas residence time.

Note that the quantities of the VOCs were not specified in these examples. Generally, if the correct gas residence time, temperature, turbulence, and excess oxygen are selected, the quantity of VOCs is not critical. The exception is very low quantities. For example, if the concentration of VOCs is less than 10 ppm, then achieving >99% destruction is more difficult. Usually, thermal oxidizer vendor specifications are written to guarantee a certain level of VOC destruction combined with a lower limit of unoxidized VOC in the stack gases.

   In selecting the design and operating conditions for a thermal oxidizer, the destruction efficiency requirements must be clearly stated. When more than one volatile organic compound is present in the waste gas, does the destruction efficiency requirement apply to each individual organic component or to the group as an aggregate?

Example 3.3: A fume stream contains methanol and toluene in equal quantities. A destruction efficiency of 97% is specified. One interpretation could be that 96% destruction of toluene and 98% destruction of methanol meets the requirements. Indeed, this is sometimes a correct interpretation. Certified EPA test methods are used in measuring stack gas emissions and for calculating VOC destruction efficiency. One method for measuring VOCs in the stack gas is EPA Method 25A. This method measures total VOCs, and does not distinguish one VOC from another. Thus, it cannot be used to calculate the destruction efficiency of a particular VOC fed to a thermal oxidizer unless it was the only VOC present in the waste stream. Other EPA test methods are available to measure the concentration of individual VOC compounds in the stack gas. An example is EPA Test Method 18.

4 Combustion Chemistry

CONTENTS

Thermal oxidation is a combustion process. Combustion is the chemical reaction of oxygen with combustible materials, accompanied by the evolution of light and the rapid production of heat. In the context of thermal oxidation, only organic compounds are considered combustible in this book. In truth, some oragnometallic compounds (e.g., hexamethylsilicane) are also combustible. These compounds are rarely encountered in thermal oxidizer applications and will not be discussed further.

An organic compound is defined as a chemical compound based on carbon chains or rings and also containing hydrogen with or without oxygen, nitrogen, or other elements. Generally, organic compounds contain only the chemical elements carbon, hydrogen, nitrogen, oxygen, chlorine, fluorine, bromine, iodine, sulfur, and phosphorus. There are exceptions, but they are not important to a discussion of thermal oxidation systems.

While the elements that comprise organic compounds are relatively few in comparison to the total known elements (106), there are more than 2 million organic compounds. In fact, the number of organic compounds outnumbers inorganic compounds. The presence of inorganic compounds in a waste stream produces products that, when thermally oxidized, usually require some type of downstream clean-up device for their removal from the combustion gas stream before it can be emitted to the atmosphere.

By definition, volatile organic compounds (VOCs) are organic compounds. A compound is a substance whose molecules consist of unlike atoms and whose constituents cannot be separated by physical means. A molecule is a group of two or more like or unlike atoms held together by chemical forces. In thermal oxidation, molecules are broken down into their simplest stable form. The reaction products from a thermal oxidation reaction are frequently termed "products of combustion" or POC.

4.1 GENERALIZED OXIDATION REACTIONS

As discussed above, thermal oxidation is the reaction of VOCs with oxygen. In simplified form:

$$VOC + O_2 \rightarrow CO_2 + H_2O + HCl + SO_2 + N_2$$

and more precisely:

$$C_aH_bN_cO_d S_eX_f + (a + e + 0.25(b - f) - 0.5d) O_2 \rightarrow aCO_2 + 0.5(b - f) H_2O$$
$$+ fHX + eSO_2 + 0.5cN_2$$

where

C	=	Carbon atom
a	=	Number of carbon atoms in the organic molecule
H	=	Hydrogen atom
b	=	Number of hydrogen atoms in the organic molecule
N	=	Nitrogen atom
c	=	Number of nitrogen atoms in the organic molecule
O	=	Oxygen atom
d	=	Number of oxygen atoms in the organic molecule
S	=	Sulfur atom
e	=	Number of sulfur atoms in the organic molecule
X	=	Any one of the halogen atoms (chlorine, fluorine, bromine, iodine)
f	=	Number of halogen atoms in the organic molecule

Shown another way:

$C \rightarrow CO_2$
$H \rightarrow H_2O$
$N \rightarrow N_2$
$S \rightarrow SO_2$ (predominantly, but also producing minor amounts of SO_3)
$Cl \rightarrow HCl$ (predominantly, but also producing minor amounts of Cl_2)

Any oxygen present in the VOC molecule itself reduces the amount of oxygen that must be added to complete the combustion reactions.

By inserting the quantity of each atom in the VOC molecule into the equation above, the combustion chemistry can be balanced. Note that the reaction as shown is with oxygen. In most cases, this oxygen is added to a thermal oxidizer as part of an air stream. Dry air consists of approximately 79% (vol) nitrogen and 21% (vol) oxygen. While the combustion reaction does not require nitrogen, it is present along with the oxygen as a consequence of using air as the source of oxygen atoms. The nitrogen is an inert gas, meaning that it does not react. For each cubic foot of oxygen required by the oxidation reaction, 3.76 ft^3 of nitrogen are added along with the oxygen.

Example 4.1: Oxidation of Methane. Methane has the chemical formula CH_4. It is oxidized as follows:

$$CH_4 + (a + e + 0.25(b - f) - 0.5\ d)\ O_2 \rightarrow aCO_2 + 0.5\ (b - f)\ H_2O$$
$$+ fHX + eSO_2 + 0.5cN_2$$

For the general equation above, a = 1, b = 4, c = 0, d = 0, e = 0, and f = 0. Thus, the coefficients of the oxidation chemistry are as follows:

$$O_2 \text{ coefficient} = (a + e + 0.25(b - f) - 0.5\ d) = (1 + 0 + 0.25(4 - 0) - 0.5(0)) = 2$$

$$CO_2 \text{ coefficient} = a = 1$$

$$H_2O \text{ coefficient} = 0.5\ (b - f) = 0.5(4 - 0) = 2$$

Therefore,

$$CH_4 + 2\ O_2 \rightarrow 1\ CO_2 + 2\ H_2O$$

Methane CH_4 was used as the combustible species in the above example. Methane is the primary constituent of natural gas and generally comprises more than 90% of the overall components of natural gas. The remaining constituents and their typical ranges are shown in Table 4.1. This analysis varies between geographic locations and can change over time.

TABLE 4.1
Typical Constituents of Natural Gas

Component	Vol%
Methane	80–95
Ethane/ethylene	3–8
Propane/propylene	0.5–3
Butanes	0.2–1
Pentanes	0–0.1
Hexane	0–0.1
Carbon dioxide	0.1–1.5
Nitrogen	0.5–10

Example: Pittsburgh, PA (1991)	
Component	**Vol%**
Methane	94.06
Nitrogen	0.28
Ethane	3.92
Butanes	0.32
Pentane	0.19
Propylene	0.87
Carbon dioxide	0.36

Natural gas is the predominant fuel used in thermal oxidation applications in the U.S. Combustion calculations usually assume that natural gas is entirely methane. This will not produce a significant error if the actual natural gas heating value is used in the energy balance calculations (discussed in Chapter 5).

Example 4.2: Thiourea — chemical formula = CH_4N_2S

$$C_aH_bN_cS_{e\,+} (a + e + 0.25(b - f))\ O_2 \rightarrow a\ CO_2 + 0.5\ (b - f)\ H_2O + f\ HCl$$
$$+ e\ SO_2 + 0.5\ c\ N_2$$

The coefficients are a = 1, b = 4, c = 2, d = 0, e = 1, and f = 0. Therefore, the balanced chemical reaction is

$$CH_4N_2S + [1 + 1 + 0.25(4 - 0)]\ O_2 \rightarrow (1)\ CO_2 + 0.5\ (4 - 0)\ H_2O$$
$$+ 0\ HCl + 1\ SO_2 + 0.5\ (2)\ N_2$$

or

$$CH_4N_2S + 3\ O_2 \rightarrow 1\ CO_2 + 2\ H_2O + 1\ SO_2 + 1\ N_2$$

In all chemical reactions, the number of atoms on each side of the equation must balance. Checking this example:

	Left side						Right Side				
	C	H	N	O	S		C	H	N	O	S
CH_4N_2S	1	4	2		1	1 CO_2	1			2	
3 O_2				6		2 H_2O		4		2	
						1 SO_2				2	1
						1 N_2			2		
Totals	1	4	2	6	1		1	4	2	6	1

As shown, there are an equal number of atoms on each side of the equation.

These reactions are shown with the exact (stoichiometric) amount of oxygen needed to complete oxidation. As will be discussed in later chapters, excess oxygen (air) is almost always added to ensure that each VOC molecule comes in contact with an oxygen molecule. Operating at low excess air levels can lead to poor VOC destruction and high carbon monoxide emissions.

4.2   HIGHLY HALOGENATED VOCS

The products of combustion of halogenated VOCs will contain some proportion of HX, where X is the halogen atom. However, hydrogen atoms must be present in sufficient quantity to ensure that HX is actually formed. In most VOC oxidation reactions, hydrogen atoms are present either as a component of the VOC or in auxiliary fuel added. However, when highly halogenated VOCs are present in high concentrations, there may be insufficient hydrogen present to convert the halide to the hydrogen halide form. Hydrogen atoms must be supplied by burning fuel that may not be needed to meet the heat balance requirements, or by water or steam injection, or by the addition of hydrogen gas.

Example 4.3: A waste stream consisting of 75% by volume phosgene and 25% nitrogen is to be thermally oxidized at 2000°F. The chemical formula for phosgene is $COCl_2$. Its lower heating value is 133 Btu/scf. Determine its oxidation chemistry.

In the absence of hydrogen atoms, the following oxidation reaction will occur:

$$COCl_2 + 1/2O_2 \rightarrow CO_2 + Cl_2$$

Ideally, it would be preferential to convert the chlorine atoms in the phosgene atom to HCl, since HCl is much easier to scrub from the products of combustion than Cl_2 This can be done by adding water vapor at high temperatures as follows:

$$COCl_2 + H_2O \rightarrow CO_2 + 2 \, HCl$$

If the concentration of phosgene in the waste gas were low, this water vapor might be produced by combustion of the auxiliary fuel required to maintain the desired operating temperature. However, the overall waste gas heating value in this example is 100 Btu/scf (0.75×133). In theory, no additional fuel is needed. Therefore, some artificial source of water vapor must be added to produce HCl rather than Cl_2.

4.3   CHEMICAL EQUILIBRIUM

The presence of certain atoms in a VOC molecule can present special problems in the design of a thermal oxidation system. Atoms falling into this category are the halogens, sulfur, and phosphorus. When organic compounds containing these atoms are oxidized, acid gases are produced. These gases are generally very corrosive to metals, and some will also attack the thermal oxidizer internal refractory lining. Their emission to the atmosphere may also be regulated.

Earlier in this chapter, the fate of each atom in an oxidation reaction was shown. In the case of acid gases, the conversions shown (e.g., $Cl \rightarrow HCl$, $S \rightarrow SO_2$) are not precisely true. While the species shown previously are almost always the predominant species, lesser amounts of related species are formed also. The relative amounts of these major and minor species are governed by chemical equilibrium. Chemical equilibrium is in turn dependent on the operating temperature of the thermal oxidizer and the concentration of the remaining components in the products of combustion. This is most easily explained using an example.

Example 4.4: A thermal oxidizer is operated at 2200°F. Combustion of a waste gas containing hydrogen sulfide produces combustion products at (essentially) atmospheric pressure, with the following composition:

	Vol%
Carbon dioxide	6.03
Water vapor	11.41
Nitrogen	77.23
Oxygen	5.31
Sulfur dioxide	0.0283

In addition to sulfur dioxide (SO_2), a small amount of sulfur trioxide (SO_3) will also be produced. The equilibrium chemistry is as follows:

$$SO_3 \leftrightarrow SO_2 + 1/2 O_2$$

The amount of each sulfur species produced is dependent upon chemical equilibrium and reaction kinetics at the conditions in the thermal oxidizer. The equation that establishes this equilibrium for SO_2/SO_3 is as follows:

$$K_{eq} = \frac{(SO_3)}{(SO_2)(O_2)^{0.5}}$$

where

K_{eq} = chemical equilibrium constant
(SO_2) = partial pressure of sulfur dioxide in combustion products
      = sulfur dioxide (vol%)/100 × total pressure (atm)
(SO_3) = concentration of sulfur trioxide in combustion products
      = sulfur trioxide (vol%)/100 × total pressure (atm)

The equilibrium constant is a function of the thermal oxidizer operating temperature. This relationship is shown graphically in Figure 4.1. It can also be determined from the following equation:[4]

$$K_{eq} = 10\wedge[\{(11996/T) - 0.362 \ln (T) + 9.36 \times 10^{-4} \times (T)$$
$$- 2.969 \times 10^5 \times (1/(T^2)) - 9.88\} /2.303]$$

where T = temperature (K)

While this equation may appear somewhat ominous, plugging in the temperature reduces it to simple mathematics. Noting that the 2200°F operating temperature is equivalent to 1477 K,

$$K_{eq} = 10\wedge[\{(11996/1477) - 0.362 \ln 1477 + 9.36 \times 10^{-4}$$
$$\times (1477) - 2.969 \times 10^5 \times 1/(1477^2) - 9.88\} /2.303]$$

$$K_{eq} = 0.0426$$

The concentration of SO_3 is then calculated as follows:

$$(SO_3) = K_{eq} (SO_2) (O_2)^{0.5}$$

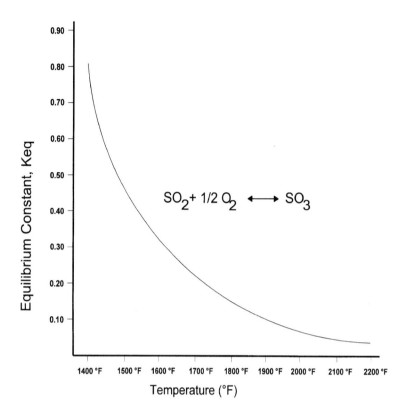

Figure 4.1 SO_2/SO_3 equilibrium constant vs. temperature.

However, when the SO_2 concentration in the combustion products was calculated, it was assumed that all of the sulfur was converted to SO_2. Therefore, the SO_2 concentration must be reduced by the amount of SO_3 formed — but this is not yet known. The problem is solved by first assuming that all sulfur is converted to SO_2, calculating the amount of SO_3 formed by using the equation above, deducting the SO_3 calculated from the SO_2, and recalculating the new SO_3 concentration from the formula above. This process is continued until the changes in SO_2 and SO_3 concentrations are insignificant. Figure 4.2 shows a logic diagram for this iterative procedure. For this example, the calculations are as follows:

Iteration Number	(SO_2)	(SO_3)
0	283.0000	2.78048
1	280.21952	2.75316
2	280.24684	2.75343
3	280.24657	2.75343
4	280.24657	2.75343

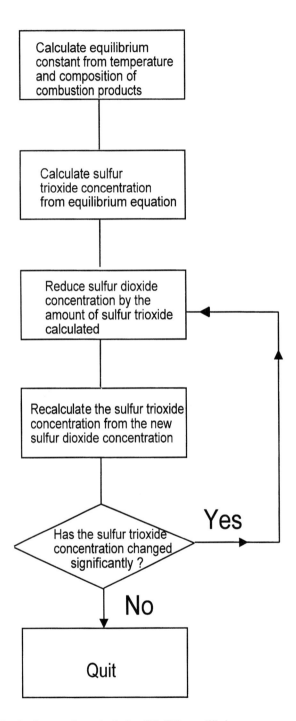

Figure 4.2   Logic diagram for calculating SO_2/SO_3 equilibrium.

Without iteration, the following formulas work just as well:

$$(SO_2, ppmv) = (SO_2)/((K_{eq} \times (O_2)^{\wedge}0.5) +1)$$

$$(SO_2, ppmv) = (0.000283)/((0.0426 \times (0.0531)^{\wedge}0.5) +1) \times 10^{\wedge}6$$

$$(SO_2, ppmv) = 280$$

$$SO_3 (ppmv) = (X - Y/10^{\wedge}6) \times 10^{\wedge}6$$

where   X   = SO_2 concentration fraction, assuming all sulfur is converted to SO_2
            = 0.000283
         Y   = SO_2 concentration (ppmv) calculated above
            = 280
SO_3 (ppmv)   = $(0.000283 - 280/10^{\wedge}6) \times 10^{\wedge}6$
SO_3 (ppmv)   = 2.8

This equilibrium distribution of sulfur oxide species applies at 2200°F. At lower temperatures, the SO_3 increases and the SO_2 decreases for the same composition of combustion products. This is shown in Table 4.2. Keep in mind, however, that in a real situation, lowering the temperature will change the composition of the combustion products. Therefore, the SO_2/SO_3 ratio will be different than shown in Table 4.2. In fact, the effect of temperature on the SO_2/SO_3 ratio is illustrated by the following example:

TABLE 4.2
SO_2/SO_3 Ratio As a Function of Temperature
Basis: Example 4.4 Combustion Products

Temperature (°F)	SO_2 (ppmv)	SO_3 (ppmv)
2200	280	2.8
1800	274	9
1400	233	49
1200	162	121

Example 4.5: Assume the waste stream is the same as in Example 4.4. However, in that example, the 2200°F operating temperature was established because the heating value of the waste gas was such that it tended to drive the temperature higher. In this example, quench air is added to lower the temperature to 1400°F. In doing so, the oxygen concentration (which affects the SO_2/SO_3 equilibrium) increases from 5.31% (vol) to 10.7%.

	Vol%
Carbon dioxide	3.96
Water vapor	7.49
Nitrogen	77.84

Oxygen	10.70
Sulfur dioxide	0.0186

At 1400°F, the value of the equilibrium constant is now 0.913. The concentrations of SO_2 and SO_3 are calculated as follows:

$$(SO_2, ppmv) = (SO_2)/((K_{eq} \times (O_2)^{0.5}) + 1)$$

$$(SO_2, ppmv) = (0.000186)/((0.913 \times (0.107)^{0.5}) + 1) \times 10^6$$

$$(SO_2, ppmv) = 143$$

$$SO_3 (ppmv) = (X - Y/10^6) \times 10^6$$

where X = SO_2 concentration fraction assuming all sulfur is converted to SO_2
            = 0.000186

         Y = SO_2 concentration (ppmv) calculated above
            = 143

SO_3 (ppmv) = $(0.000186 - 143/10^6) \times 10^6$
SO_3 (ppmv) = 43

In Example 4.5, the ratio of SO_2/SO_3 is 143/43 = 3.33, whereas the ratio in Table 4.2 is 233/49 (= 4.76) at the same 1400°F temperature. This illustrates the need to establish the composition of the combustion products at different temperatures, rather than just to use the same composition at different temperatures to determine the effect of temperature on the SO_2/SO_3 equilibrium ratio.

This example assumed that chemical equilibrium alone determined the final reaction products. In real situations, chemical kinetics may govern. In fact, with SO_2/SO_3 containing gases at temperatures less than 1800°F the concentration of SO_3 is governed by chemical kinetics and usually does not exceed 5% of the total SO_2/SO_3 present. Therefore, it is normally assumed that the maximum SO_3 concentration is set by the equilibrium constant at 1800°F.

The following example demonstrates the equilibrium calculations for a chlorinated waste.

Example 4.6: A waste stream containing allyl chloride is injected into a thermal oxidizer for destruction. The thermal oxidizer is operated at 1800°F and ambient pressure. The combustion products are as follows:

	Vol%
Carbon dioxide	5.09
Water vapor	12.17
Nitrogen	69.94
Oxygen	11.64
Hydrogen chloride	1.16

In addition to hydrogen chloride (HCl), a small amount of chlorine gas (Cl_2) will also be produced. The equilibrium chemistry is as follows:

$$2 \, HCl + 1/2O_2 \leftrightarrow Cl_2 + H_2O$$

$$K_{eq} = \frac{(Cl_2)(H_2O)}{(HCl)^2(O_2)^{0.5}}$$

The iterative procedure described in Figure 4.2 must be used. In this case, each chlorine atom produced consumes two HCl atoms. The equilibrium constant can be calculated from the following formula:

$$K_{eq} = 10\char`\^(3114.7/T(°K) - 3.5816)$$

$$\text{At } 1800°F \, (1255 \, K), \, K_{eq} = 0.079$$

The problem is solved by first assuming that all chlorine in the allyl chloride is converted to HCl, calculating the amount of chlorine gas (Cl_2) formed by using the equation above, deducting the resulting chlorine gas concentration calculated from the previously calculated HCl concentration, and recalculating the new chlorine gas concentration from the formula above. This process is continued until the change in HCl and Cl_2 concentrations are insignificant. The iterations are shown as follows:

Iteration Number	(HCl)	(Cl_2)
0	11,623	30.1
1	11,563	2.8
2	11,563	2.8

Since the Cl_2 concentration is not changing, no further iteration is required.

HCl is easily scrubbed from combustion products using a water scrubber (discussed in Chapter 13). However, chlorine gas cannot be scrubbed with water alone. Raising the thermal oxidizer operating temperature will increase the HCl/Cl_2 ratio (i.e., more HCl, less chlorine gas). This method is sometimes used to reduce the Cl_2 concentration to the point where only a water scrubber is needed to meet emission limitations.

The effect of temperature on the equilibrium constant is shown in Figure 4.3. The larger the equilibrium constant, the more chlorine gas produced. In the example above, assuming the composition of the waste gas does not change with temperature, the ratio of HCl/Cl_2 is 913 at 2200°F and 115 at 1400°F.

Equilibrium constants as a function of temperature are summarized in Table 4.3 for other halogens of interest, including HBr/Br_2, HF/F_2, and HI/I_2. For iodine, virtually all of the halogen is converted to the I_2 form. With bromine, more Br_2 is normally produced as opposed to HBr. In contrast, almost all fluorine is converted to the HF form.

4.4 DEWPOINT

Halogenated gases and SO_2/SO_3 become corrosive to metals when they reach their dewpoint. The dewpoint is the temperature at which liquid condensation of the HX

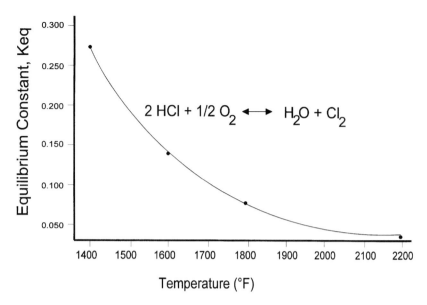

Figure 4.3   HCl/Cl$_2$ equilibrium constant vs. temperature.

TABLE 4.3
Equilibrium Constant Equations

For X = Halogen (Cl, F, Br, I)
$$2\ HX + 1/2\ O_2 \leftrightarrow X_2 + H_2O$$

$$K_{eq} = \frac{(X_2)(H_2O)}{(HX)^2(O_2)^{0.5}}$$

For HCl/Cl$_2$, $K_{eq} = 10^{\wedge}(3114.7/T(K) - 3.5816)$
For HF/F$_2$, $K_{eq} = 1/(2.718^{\wedge}(5.874 \times ((T(K)/6672.923)^{\wedge}(-0.98) \times 2.718^{\wedge}(T(°K))/6672.923)))$
For HBr/Br$_2$, $K_{eq} = 1707.97 \times (T(°F)^{\wedge}(-3010.123/(T(°F)))$
For HI/I$_2$, $K_{eq} = 10^{\wedge}(15.216 \times (T(K))^{\wedge}2 + 321.423 \times (T(K)) + 2645.6)$

Source: Derived from Cudahy, J.J., Eicher, A.R., and Troxler, W.L., Thermodynamic equilibrium of halogen and hydrogen halide during the combustion of halogenated organics," 6th Nat. Conf. Management of Uncontrolled Hazardous Waste Sites, Washington, D.C., November 1985.

SO$_2$/SO$_3$ SO$_2$ + 1/2 O$_2$ $\leftrightarrow$ SO$_3$

$$K_{eq} = \frac{(SO_3)}{(SO_2)(O_2)^{0.5}}$$

$K_{eq} = 10^{\wedge}[\{11996/T - 0.362 \ln T + 9.36 \times 10^{\wedge}{-4} \times T - 2.969 \times 10^5 \times (1/T^2) - 9.88\}/2.303]$
where T = temperature (K)

Source: Reynolds, J.P., Dupont, R.R., Theodore, and L. *Hazardous Waste Incineration Calculations: Problems and Software*, John Wiley & Sons, New York, 1991.

and SO_3 (as H_2SO_4) begins to form. For example, HCl formed can combine with water vapor in the products of combustion, condense on a cold oxidizer inner metal shell, and initiate corrosion. This can be prevented by operating with a relatively hot shell (>300°F). Since this temperature is above the dewpoint temperature of HCl, condensation cannot occur. The hot shell temperature is achieved by selecting a less insulating inner refractory material, reducing the amount of refractory material, or externally insulating the thermal oxidizer outer shell. These same considerations should be applied to ducting, boiler economizer tubes, and stacks, if associated with a thermal oxidizer system destroying halogenated compounds. Hot external surfaces should be shielded for personnel protection.

As described earlier, the acidic forms of concern are

- Sulfuric acid (H_2SO_4) from SO_2 /SO_3
- Hydrogen chloride from HCl/Cl_2
- Hydrogen bromide from HBr/Br_2
- Hydrogen iodide from HI/I_2
- Hydrofluoric acid from HF/F_2

Equations for calculating the dewpoint of these species are shown in Table 4.4. The water vapor content of the combustion products affects the dewpoint in all cases. Generally, if the water vapor content is higher, the dewpoint is higher. However, the concentration of the acid species has a greater impact on the dewpoint than the water vapor content.

Ensuring that all surfaces that come in contact with the combustion products are above these dewpoint temperatures prevents dewpoint corrosion. Since the equations shown in Table 4.4 are not the usual English units, an example will illustrate their use:

Example 4.7: Combustion of a waste gas containing hydrogen sulfide produces combustion products at atmospheric pressure with the following composition:

	Vol%
Carbon dioxide	6.03
Water vapor	11.41
Nitrogen	77.23
Oxygen	5.31
Sulfur dioxide	0.0283
Sulfur trioxide	0.0003 (by chemical equilibrium)

Calculate the sulfuric acid dewpoint.

Using the equation for sulfuric acid in Table 4.4:

$$1000/T_{dp} = 1.7842 + 0.0269 \log(P_{H_2O}) - 0.1029 \log(P_{SO_3})$$
$$+ 0.0329 \log(P_{H_2O})\log(P_{SO_3})$$

Atmospheric pressure = 1 atmosphere

TABLE 4.4
Acid Gas Dewpoints

Hydrobromic acid (HBr)

$$1000/T_{dp} = 3.5639 - 0.135 \ln (P_{H_2O}) - 0.0398 \ln (P_{HBr}) + 0.00235 \ln (P_{H_2O}) \ln (P_{HBr})$$

Hydrochloric acid (HCl)

$$1000/T_{dp} = 3.7358 - 0.1591 \ln (P_{H_2O}) - 0.0326 \ln (P_{HCl}) + 0.00269 \ln (P_{H_2O}) \ln (P_{HCl})$$

Hydrofluoric acid (HF)

$$1000/T_{dp} = 3.8503 - 0.1728 \ln (P_{H_2O}) - 0.02398 \ln (P_{HF}) + 0.001135 \ln (P_{H_2O}) \ln (P_{HF})$$

where T_{DP} = dewpoint temperature (K)
　　　P = total pressure (mmHg) × volume fraction of component in POC

Source: Kiang, Y.H., Predicting dewpoints of acid gases, *Chem. Eng.* (February 9, 1981).

Sulfuric acid (H_2SO_4)

$$1000/T_{dp} = 1.7842 + 0.0269 \log(P_{H_2O}) - 0.1029 \log(P_{SO_3}) + 0.0329 \log(P_{H_2O}) \log(P_{SO_3})$$

where T_{DP} = dewpoint temperature (K)
　　　P = partial pressure (atm)

Source: Pierce, R.R., *Chem. Eng.*, April 11, 1977.

$$P_{H_2O} = (11.41/100) = 0.1141 \text{ atm}$$

$$(P_{SO_3} = (0.003/100) = 0.000003 \text{ atm} = 3 \times 10^{-6})$$

Substituting into the dewpoint equation:

$$1000/T_{dp} = 1.7842 + 0.0269 \log(P_{H_2O}) - 0.1029 \log(P_{SO_3})$$
$$+ 0.0329 \log(P_{H_2O}) \log(P_{SO_3})$$

$$1000/T_{dp} = 1.7842 + 0.0269 \log (0.1141) - 0.1029 \log (3 \times 10^{-6})$$
$$+ 0.0329 \log (0.1141) \log (3 \times 10^{-6})$$

$$T_{dp} = 400 \text{ K} = 127°C = 261°F$$

4.5　PRODUCTS OF INCOMPLETE COMBUSTION (PICs)

While acid gases are undesirable because of their corrosive affects, another class of undesirable combustion products is called "products of incomplete combustion" (PIC). When there is insufficient temperature, time, turbulence, or oxygen in a thermal oxidizer, a VOC may not be oxidized completely. While it may no longer exist in its original form, it may form other organic compounds in addition to carbon dioxide and water vapor. Typical PICs are aldehydes, ketones, and ethers. Aldehydes can contribute to odor problems from a stack. Consequently from an environmental standpoint, the thermal oxidizer has not achieved its purpose of completely destroying all organic compounds. Typically, these PICs are detected using standard EPA stack gas measurement techniques and are counted against the overall VOC destruction efficiency. Fortunately, a thermal oxidizer that is designed and operated properly will have very low emissions of PICs. Carbon monoxide emissions are usually a key indicator of incomplete combustion. If CO emissions are low (e.g., < 50 ppmv), PIC formation is normally minimal.

4.6 SUBSTOICHIOMETRIC COMBUSTION

Partial oxidation occurs when substoichiometric (less than theoretical) amounts of oxygen are added to a combustion process. Here, the oxygen supplied to the combustion system is insufficient for combustion of all the organic compounds present. There are times when this may be done purposely. For example, as described in more detail in Chapter 11, it may be beneficial to first partially oxidize any waste gases containing nitrogen species. In this way, most of the nitrogen is converted to innocuous molecular nitrogen gas (N_2) rather than NOx, a pollutant itself. The partial oxidation is followed by injection of additional air/oxygen in a downstream section of the thermal oxidizer to oxidize the partial combustion products (e.g., CO, H_2).

This partial oxidation of the organic compounds leads to a more complex chemistry. For example, the carbon present in the organic compounds is converted to both carbon monoxide and carbon dioxide. The relative quantities of each are determined by chemical equilibrium, as are the halogen and sulfur acid gases. The reactions occurring are as follows:

$$2\ C + 3/2\ O_2 \leftrightarrow CO + CO_2$$

$$4\ H + 1/2\ O_2 \leftrightarrow H_2 + H_2O$$

From these species, additional equilibrium reactions are generated as follows:

$$CO + H_2O \leftrightarrow H_2 + CO_2$$

The above reaction is known as the water–gas shift reaction and is well known in other areas of combustion and gasification. The complex equlibrium in partial oxidation is better left to computer codes for its solution. But as with the acid–gas equilibrium, it is a function of operating temperature.

4.7 EMISSION CORRECTION FACTORS

The design of a thermal oxidizer is almost always guided by the requirement to achieve emission levels specified in an environmental permit. These emission levels, on a volume concentration basis, are affected to some extent by the excess air (oxygen content of the combustion products) used in the thermal oxidizer. Increasing the excess air increases the vloume of the combustion products and dilutes pollutant species in the combustion products. At one time there was a saying: "the solution to pollution is dilution." However, the EPA recognized that this approach did not meet the intent of the regulations. Therefore, many times emissions levels (e.g., NOx, CO, SO$_2$) must be corrected to a specific level of oxygen or carbon dioxide in the stack gas. Typical values are 3 and 7% O$_2$ (dry basis) and 12% CO$_2$. If the correction is on a dry basis, the oxygen concentration of the combustion products must first be adjusted to a dry basis concentration before applying the correction factor. In general the correction factor applies as follows for oxygen:

$$C_f = \frac{(21 - \text{reference } \% \ O_2)}{(21 - \text{measured } \% \ O_2)}$$

where C$_f$ = correction factor

For CO$_2$, the correction factor is a direct ratio.

$$C_f = \frac{\text{measured } \% \ CO_2}{\text{reference } \% \ CO_2}$$

In both cases, the measured emission value is multiplied by the correction factor.

Example 4.8: A stack gas has the following composition. Correct the NOx emissions to 3% oxygen on a dry basis:

Component	Vol%
Carbon dioxide	4.8
Water vapor	6.4
Nitrogen	82.9
Oxygen	5.9
NOx	0.0025 (dry basis)

(*Note*: Volume % is converted to ppmv by multiplying by 10,000. In the above example, 0.0025 vol% NOx = 10,000 × 0.0025 = 25 ppmv). Most analytical instruments measure NOx on a dry basis.

The oxygen content of the stack gas must be first converted to a dry basis as follows:

$$\text{Dry } \% \ O_2 = \text{Wet } \% \ O_2 / (100 - \% \text{ water vapor})$$

$$\text{Dry } \% \ O_2 = 5.9 / (100 - 6.4) = 6.30$$

Therefore the correction factor is

$$C_f = (21 - 3)/(21 - 6.3) = 1.224$$

The corrected NOx emission is the measured amount multiplied by the correction factor. In this case:

$$\text{Corrected NOx (dry @ 3\% O}_2) = 25 \times 1.224 = 31 \text{ ppmv}$$

There are situations where the oxygen correction factor must be applied, even when there has been no attempt to dilute the combustion products. For example, the oxygen content of a VOC-contaminated air stream remains very high during the thermal oxidation process. It may still be in the range of 18 to 20% in the stack gas, even though no additional air whatsoever was added to the thermal oxidizer. Regulators should recognize that in these cases there has been no attempt to dilute the combustion products and the oxygen correction should not be applied.

In the U.S., stack emissions of pollutant species are typically reported on a ppmv basis. However, in Europe and occasionally in the U.S., emissions are reported as mg/Nm^3 (milligrams per normal cubic meter). Normal conditions are equivalent to standard conditions in the U.S. except that normal temperature is 32°F (0°C). These two sets of units can be converted as follows:

$$mg/Nm^3 \times 22.38/MW = ppmv$$

Example 4.9: Convert 100 ppmv of NOx (as NO_2) and 50 ppmv of carbon monoxide to mg/Nm^3.

The molecular weight of NOx (as NO_2) is 46 and 28 for carbon monoxide. Therefore

$$100 \text{ ppmv NO}_2/ (22.38/46) = 206 \text{ mg/Nm}^3$$

$$50 \text{ ppmv CO}/(22.38/28) = 63 \text{ mg/Nm}^3$$

4.7.1 WEIGHT PERCENT TO VOLUME PERCENT CONVERSION

In thermal oxidation systems, the composition of both the waste stream and combustion products is almost always specified in terms of volume percentage (%). However, volume percents can be converted to weight percents using the following formulas:

$$W = V \times (MWv/Mwo)$$

where

    W   = Wt% of gas component
    V   = Vol% of gas component
    MWv= molecular weight of the gas reported as vol%
    Mwo = overall molecular weight of total gas stream

Example 4.10: The combustion products from a thermal oxidizer in volume percent are as follows. Convert each component to weight percent.

Component	Vol%
Carbon dioxide	5.41
Water vapor	54.07
Nitrogen	37.53
Oxygen	3.00

The overall molecular weight of this gas mixture is ($0.0541 \times 44 + 0.5407 \times 18.01 + 0.3753 \times 28.01 + 0.03 \times 32$) = 23.6. Converting to weight percent:

Component	Vol%	Wt%
Carbon dioxide	5.41	$0.0541 \times 44.01/23.6 = 10.09$
Water vapor	54.07	$0.5407 \times 18.02/23.6 = 41.29$
Nitrogen	37.53	$0.3753 \times 28.01/23.6 = 44.54$
Oxygen	3.00	$0.03 \times 32.0/23.6 = 4.07$

5 Mass and Energy Balances

CONTENTS

5.1 FUNDAMENTALS

A thermal oxidation system raises the temperature of a VOC in a waste gas stream to a level where oxidation of the VOC occurs. In the process, air, auxiliary fuel, and occasionally water or steam may be added; but, in all cases, the mass of the input streams must be exactly equivalent to the total mass of the output streams. In general, for all chemical reactors:

$$\text{Input (mass)} = \text{Output (mass)} + \text{accumulation}$$

In a thermal oxidizer, there is no accumulation term. Therefore, input must equal output.

Let's take the simple example of combustion of natural gas. The reactants are shown in the following equation:

$$CH_4 + O_2 \rightarrow CO_2 + H_2O$$

But the reaction is not balanced. There must be an equal number of atoms of like kind on each side of the equation. Using the formula for balancing combustion reactions shown in Chapter 4 results in the following balanced equation:

$$CH_4 + 2\ O_2 \rightarrow CO_2 + 2\ H_2O \text{ (with oxygen)}$$

$$CH_4 + 2\ O_2 + 7.6\ N_2 \rightarrow CO_2 + 2\ H_2O + 7.6\ N_2$$
$$\text{(with air as the source of oxygen)}$$

Since there are equal numbers of atoms of like kind equal on both sides of this equation, the weights (mass) on each side of the equation are equal. When the exact amount of oxygen is added, based on the balanced equation above, this quantity is called the stoichiometric amount. In most combustion reactions, excess oxygen (or air) is added. Excess air rates will be discussed further later in this chapter.

5.1.1 MOLECULAR WEIGHT

To demonstrate that there is equal mass on both sides of the balanced equation, we must first calculate the molecular weight of each reacting species. The molecular weight of any compound is the sum of the number of atoms of a particular element multiplied by its atomic weight. Each element has a specific atomic weight. The atomic weights of all known elements are shown in Appendix B. Fortunately, in thermal oxidation systems, only a few of the total number of existing elements are normally present. The atomic weights of the elements commonly encountered in thermal oxidation reactions are as follows:

Element	Atomic Symbol	Atomic Weight
Carbon	C	12.01
Hydrogen	H	1.01
Nitrogen	N	14.01
Oxygen	O	16.00
Sulfur	S	32.06
Chlorine	Cl	35.45

The molecular weights of the substances participating in the methane reaction shown above are as follows:

$$CH_4 = 12.01 + 4 \times 1.01 = 16.05$$

$$O_2 = 2 \times 16.00 = 32.00$$

$$CO_2 = 12.01 + 2 \times 16.00 = 44.01$$

$$H_2O = 2 \times 1.01 + 16.00 = 18.02$$

5.1.2 MOLES

The mass of a chemical compound divided by its molecular weight equals the number of moles of that substance. In the methane reaction, 1 mole of methane combines with 2 moles of oxygen to form 1 mole of carbon dioxide and 2 moles of water vapor. Although the total number of moles is equal on each side of the methane combustion reaction, this is coincidental. In general, the number of moles do not need to balance, only the mass.

If the mass is specified in pounds, the number of moles is termed "pound moles" and designated by the abbreviation lb-mol. Each lb-mol of a chemical compound has a volume of 379 ft³ at standard conditions. As used here, standard conditions are 60°F and 1 atmosphere pressure. However, standard conditions are anything but standard. Different industries define different temperatures as standard. The various "standard" temperatures are 32°F (0°C), 60°F, 68°F, 72°F, and 77°F. In this book, 60°F is used as the standard temperature for flow calculations and 77°F for heats of combustion. The error produced by this inconsistency is negligible.

While chemical engineers and chemists may prefer to work with pound mole (lb-mol), other disciplines may be more comfortable using standard cubic feet (scf). They are convertible using the following formulas:

$$scf = lb\text{-}mol \times 379$$

$$lb\text{-}mol = lb/MW$$

$$scf = lb/MW \times 379$$

$$mole\ \% = volume\ \%$$

$$where\ scf = standard\ cubic\ feet$$

$$MW = molecular\ weight$$

Since scf are convertible to lb-mol, they can be used in the same way in balancing a thermal oxidation reaction. Using the methane combustion example with 10 scfm as the basis:

	CH_4 +	$2\ O_2$ →	CO_2 +	$2\ H_2O$
scfm	10	20	10	20
lb-mol/min	10/379	20/379	10/379	20/379
lb/min	(10/379) × 16.04	(20/379) × 32	(10/379) × 44.01	(20/379) × 18.02

Reducing these equations results in the following:

	CH_4 +	$2\ O_2$ →	CO_2 +	$2\ H_2O$
scfm	10	20	10	20
lb-mol/min	0.0264	0.053	0.0264	0.053
lb/min	0.423	1.696	1.161	0.955

$$Total\ mass\ input = 0.423 + 1.696 = 2.12\ lb/min$$

$$Total\ mass\ output = 1.161 + 0.955 = 2.12\ lb/min$$

Again, the fact that the lb-mol/min and scfm balance is purely coincidental, as illustrated in Example 5.1.

Example 5.1: (25 scfm) Benzene is oxidized with oxygen gas. Determine the combustion products. Benzene has the chemical formula of C_6H_6 and thus a molecular weight of 78.12

	C_6H_6 +	$15/2\ O_2$ →	$6\ CO_2$ +	$3\ H_2O$
scfm	25	$15/2 \times 25$	6×25	3×25
lb-mol/min	25/379	$15/2 \times 25/379$	$6 \times 25/379$	$3 \times 25/379$
lb/min	$25/379 \times 78.12$	$15/2 \times 25/379 \times 32.00$	$6 \times 25/379 \times 44.01$	$3 \times 25/379 \times 18.02$

or

	C_6H_6 +	$15/2\ O_2$ →	$6\ CO_2$ +	$3\ H_2O$
scfm	25	187.5	150	75
lb-mol/min	0.066	0.495	0.396	0.198
lb/min	5.16	15.84	17.43	3.57

Total mass in = 5.16 + 15.84 = 21.0

Total mass out = 17.43 + 3.57 = 21.0

Note in this example that the scfm and lb-mol/min are not equal on both sides of the equation.

5.2   ENERGY BALANCE

While the mass in and mass out must be equal, so must the energy. Since there is no mechanical work performed by a thermal oxidizer, this energy balance is stated in terms of an enthalpy balance. Technically, enthalpy is defined as the internal energy plus the product of the pressure and the volume. Alternately, it is heat minus work. As applied to thermal oxidizer energy calculations, it is the heat content of a substance relative to a reference state (temperature and pressure). It is a product of the mass of the substance, the heat capacity of the substance, and the difference between its actual temperature and a standard reference temperature. In English units, heat is described in terms of British thermal units (Btu). A Btu is the amount of heat required to raise the temperature of 1 pound of water by 1°F. Heat capacity is the amount of heat required to increase the temperature of any substance by 1°F. It is only identical to one Btu when referring to water. Heat capacity varies for different substances and is a function of temperature. Its relationship with temperature will be discussed later in this chapter. In general:

$$\delta H = Q$$

where   Q = heat and δH = change in enthalpy

Expanding this equation:

$$\delta H = Q = m\ Cp\ (T_a - T_r)$$

where

      m   = mass of substance (lb)
      Cp = mean heat capacity of that substance (Btu/lb-°F)
      T_a = temperature of the substance (°F)
      T_r = reference temperature (°F), such as 77°F (25°C)

This equation applies to solids, liquids, and vapors.

Example 5.2: The temperature of 10 lb of liquid water is raised from 32 to 212°F (but the water remains a liquid). The mean heat capacity of water over this temperature range is 1.0 Btu/lb-°F. Using 32°F as the reference temperature (enthalpy = 0), then the enthalpy change is

$$\delta H = Q = m\ Cp\ (T_a - T_r)$$

$$\delta H = 10\ lb \times 1.0\ Btu/lb\text{-}°F \times (212 - 32)$$

$$\delta H = 1800\ Btu$$

This matches the value given in steam tables that list the enthalpy of water as 180 Btu/lb at 212°F.

The change in the heat content of a substance without a change in physical state is termed "sensible heat." When there is a change in state, the heat required (or evolved) is called "latent heat." For example, 970 Btu/lb (at atmospheric pressure) of heat must be added to evaporate the water heated to 212°F in the above example. Therefore, 10 lb of water requires 9700 Btu. The latent heat required to evaporate the water exceeds the sensible heat required to raise the temperature of the water from 32 to 212°F. To raise the temperature of the water vapor even further requires sensible heat again as follows:

$$\delta H = Q = m\ Cp\ (T_f - T_i)$$

where T_f and T_i are the final and initial temperatures, respectively. The heat capacity in this case is the mean heat capacity for the vapor between the initial and final temperatures. Thus, to raise the temperature of 10 lb of water vapor to, say, 1500°F, the heat required is

$$Q = 10\ lbs. \times 0.495\ Btu/lb\text{-}°F \times (1500°F - 212°F)$$

$$Q = 6376\ Btu$$

where 0.495 Btu/lb-°F is the mean heat capacity of water vapor.

Thus, to raise the temperature of 10 lb of liquid water (and convert to water vapor) from 32 to 1500°F requires a total of

$$1800 + 9700 + 6376 = 17876 \text{ Btu (or 1788 Btu/lb)}$$

TABLE 5.1
Mean Molar Heat Capacities for Typical Combustion Gases (Btu/lb-mol-°F)

Temperature (°F)	N_2	O_2	Air	CO_2	H_2O	HCl	SO_2
600	7.02	7.26	7.07	9.97	8.25	7.05	10.6
800	7.08	7.39	7.15	10.37	8.39	7.10	10.9
1000	7.15	7.51	7.23	10.72	8.54	7.15	11.2
1200	7.23	7.62	7.31	11.02	8.69	7.19	11.4
1400	7.31	7.71	7.39	11.29	8.85	7.24	11.7
1600	7.39	7.80	7.48	11.53	9.01	7.29	11.8
1800	7.46	7.88	7.55	11.75	9.17	7.33	12.0
2000	7.53	7.96	7.62	11.94	9.33	7.38	12.1
2200	7.60	8.02	7.69	12.12	9.48	7.43	12.2
2400	7.66	8.08	7.75	12.28	9.64	7.47	12.3

Source: Williams, E.T. and Johnson, R.C., *Stoichiometry for Chemical Engineers*, McGraw Hill, New York, 1958. With permission.

The concept of enthalpy and heat is relatively straightforward. However, the variation of heat capacity (Cp) with temperature must be known. That variation for air and typical combustion products (in units of Btu/lb-mol-°F) is presented in tabular form in Table 5.1. To convert from Btu/lb-mol-°F to Btu/lb-°F, divide the values shown by the molecular weight of the gas. The molecular weights of the gases shown are as follows:

Nitrogen	28.01
Oxygen	31.99
Carbon dioxide	44.01
Water vapor	28.01
Air	28.85
Hydrogen chloride	36.45
Sulfur dioxide	64.06

Equations for calculating heat capacities as a function of temperature for these common combustion products are shown in Table 5.2.

For rigorous calculations, the heat capacity of each VOC injected into the thermal oxidizer must also be known. The heat already contained in the waste stream is calculated and deducted from the heat necessary to thermally oxidize the VOCs and raise the combustion products to their final temperature. This concept will be elaborated further later in this chapter.

TABLE 5.2
Mean Molar Heat Capacities for Typical Combustion Gases (Btu/lb-mol-°F)

Heat capacity $= A \times [(2.71828^{((Ln(T) - B)^2)/C})]$
T = Temperature (°F)

Gas	A	B	C
CO_2	7.369	1.324	82.591
H_2O	8.035	5.701	23.727
N_2	6.941	5.696	45.022
O_2	6.953	4.576	69.314

For HCl

$$\text{Heat capacity} = \frac{[((T^3 \times (-1.036 \times 10^{-9}) - (0.317 \times 10^{-5}) \times T^2 - (0.182 \times 10^{-2}) \times T + 7.244) + 6.96)]}{2}$$

For SO_2

$$\text{Heat capacity} = \frac{[((T^3 \times (2.057 \times 10^{-9}) - (0.9103 \times 10^{-5}) \times T^2 + (0.134 \times 10^{-2}) \times T + 6.157) + 9.54)]}{2}$$

where T = temperature (K)

Source: Hougen, O., Watson, K., and Ragatz, R., *Chemical Process Principles*, John Wiley & Sons, New York, 1964. With permission.

There are two exceptions when the heat capacity of the VOCs is unimportant: (1) when the waste gas is at ambient temperature and contains no sensible heat (compared to the reference temperature) and (2) when the concentration of VOC is so small as to have a negligible effect on the heat balance. With regard to the first exception, the enthalpy of the waste gas is a function of the difference between the waste stream temperature and a standard reference temperature. If, for example, 77°F (25°C) is selected as the reference temperature and the waste stream temperature is at a temperature of 77°F, then the enthalpy of the waste stream is zero. The enthalpy is also a function of the mass of a component. With reference to the second exception, if the mass of VOCs are negligible compared to the total mass of the waste stream, then they will have a negligible affect on the heat balance even if they are at an elevated temperature.

5.3 LOWER AND HIGHER HEATING VALUES

Often the mass of VOCs themselves is small compared to the total mass of a waste stream (containing mostly inert components such as nitrogen). Nonetheless, the heat released from oxidation of this relatively small mass can still be significant. Almost every organic compound releases a specific quantity of heat when oxidized. The heat release from thermal oxidation of a VOC containing waste stream can be substantial and in many cases is adequate to reach the operating temperature of the thermal oxidizer without the addition of auxiliary fuel. This heat release is called the heat of combustion and in English units is typically specified in terms of Btu/lb. It can be specified as higher heating value (HHV) or lower heating value (LHV).

The higher heating value is sometimes called the "gross" heating value and the lower heating value the "net" heating value. The user of heating value data must be careful to understand the difference and use the correct value in calculations. These two terms represent the same information but in a different format. The higher heating value represents oxidation with stoichiometric air at a specified reference temperature with the final combustion products at the same temperature and any water produced in the liquid state. The lower heating value has the same definition, except that any water produced is in the vapor state. Most thermodynamic data are listed at a reference temperature of 25°C (77°F). That will be the reference temperature used in this book for combustion calculations.

To illustrate the difference between higher and lower heating values, combustion of methane will be used as an example. Its reaction is shown below.

$$CH_4 + 2\ O_2 \rightarrow CO_2 + 2\ H_2O$$

The difference between higher and lower heating value is due to the water produced. Heat is required to evaporate water (1050 Btu/lb at 77°F and atmospheric pressure). (The heat of vaporization of water does vary with temperature. That explains why the value just given above at 77°F [1050 Btu/lb] does not match the value given earlier in the chapter [970 Btu/lb], which was at a reference temperature of 212°F.) Conversely, when water vapor is condensed it releases heat.

The tabulated value for the HHV of methane is 23,879 Btu/lb. This is the total heat released per pound of methane combusted, assuming that the water vapor produced is condensed and its final form is liquid. In contrast, the lower heating value assumes that the water remains a vapor. The tabulated value for the LHV of methane is 21,520 Btu/lb. Again, the difference is the phase (liquid vs. vapor) of the water. Since the reaction chemistry is known, the value in one basis can be converted to the other. Since 2 moles (scfm) of water vapor are formed for each mole (scfm) of methane combusted, the difference between the HHV and LHV is calculated as follows:

$$LHV \ = \ HHV - \frac{(N \times 18.02)}{MW_{VOC}} \times 1050.2$$

where
        LHV   = lower heating value (Btu/lb)
        HHV   = higher heating value (Btu/lb)
        N   = number of moles of water vapor produced per mole of VOC
        MW_{VOC} = molecular weight of the VOC or organic compound

Assuming 1 mole of methane initially:

$$N = 2 \text{ moles of water vapor produced per mole of methane}$$

$$MW_{VOC} = 16.04 \text{ (methane)}$$

$$\text{HHV} = 23{,}879 \text{ Btu/lb of methane}$$

$$\text{LHV} = 23{,}879 - \frac{(2 \times 18.02)}{16.04} \times 1050.2 = 21{,}519 \text{ Btu/lb}$$

In the same way, the LHV can be converted to the HHV by rearranging the formula above.

In thermal oxidation systems, the water vapor produced from the combustion reactions is rarely condensed. Thus, the LHV is normally used in heat balance calculations. However, some tabulated data lists heats of combustion as HHV. If so, the formula above can be used to convert HHV to LHV.

Even if the heat released by a VOC is sufficient to raise its temperature to the thermal oxidizer operating temperature, the oxidation reaction must be initiated. The concept of autoignition temperature (AIT) was introduced in Chapter 4. To release the heat contained in a VOC, the temperature of the VOC must be first raised to its AIT. Therefore, an auxiliary fuel burner is usually required, at least for start-up, regardless of the heat content of the waste stream itself.

5.4   AUXILIARY FUELS

In many cases, the heat released by combustion of a VOC is insufficient to raise the temperature of the waste stream (plus any combustion air needed) to the thermal oxidizer operating temperature. Generally, in the U.S., natural gas is added as a supplemental fuel to generate additional combustion heat needed to achieve the target operating temperature. However, fuel oil is also occasionally used. The heating value of fuel oil is usually listed in terms of HHV. Since fuel oil consists of not one but many complex organic compounds, rigorous calculations to convert HHV to LHV are more complex. Usually, a value of 90% of the HHV is a close approximation to the LHV.

5.5   MASS-TO-VOLUME HEAT RELEASE CONVERSIONS

While the previous discussion focused on heat released from organic or VOCs in terms of heat per pound, it is just as common to use units of heat per standard cubic foot (Btu/scf). Again, knowing the molecular weight of the heat-releasing substance, conversion from Btu/lb to Btu/scf can be performed using the following formula:

$$\text{Btu/scf} = \text{Btu/lb} \times \text{MWvoc}/379$$

where
$$\text{MWvoc} = \text{molecular weight of VOC}$$

Again, using methane as the example:

$$\text{Btu/scf} = 21{,}520 \text{ Btu/lb (LHV)} \times 16.04/379$$

$$\text{Btu/scf} = 911$$

This tabulated value for methane is 913 Btu/scf. The concept of lower and higher heating values applies to these units (Btu/scf), also.

5.6 MIXTURE HEATING VALUES

Rarely is a waste stream composed of a single component. It is typically a mixture of VOCs and air or inert gases (e.g., N_2, CO_2). The overall heating value of the waste stream can be calculated by multiplying the volume percent of each combustible component by its heating value. The following example illustrates these calculations.

Example 5.3: A coke oven off-gas has the following composition. Determine its heating value.

Component	Vol%
Carbon monoxide	4.5
Hydrogen	57.9
Methane	30.3
Nitrogen	5.5
Carbon dioxide	1.8

The heating values of each of these components are as follows (LHV):

Component	LHV (Btu/scf)
Carbon monoxide	322
Hydrogen	275
Methane	913
Nitrogen	0
Carbon dioxide	0

The contribution of each component to the overall heating value is as follows:

Heat Released

Component	LHV (Btu/scf)
Carbon monoxide	$0.045 \times 322 = 14.5$
Hydrogen	$0.579 \times 275 = 159.2$
Methane	$0.303 \times 913 = 276.6$
Nitrogen	$0.055 \times 0 = 0$
Carbon dioxide	$0.018 \times 0 = 0$

The total is 450 Btu/scf.

Heating values of some common fuels, hydrocarbons, and other combustible compounds are shown in Table 5.3. Appendix C contains a more complete listing.

TABLE 5.3
Heats of Combustion of Common Combustible Gases

Compound	Heat of Combustion		Compound	Heat of Combustion	
	LHV (Btu/lb)	LHV (Btu/scf)		LHV (Btu/lb)	LHV (Btu/scf)
Hydrogen	51,623	275	Ethylene	20,295	1,513
Carbon monoxide	4,347	322	Propylene	19,691	2,186
Methane	21,520	913	Benzene	17,480	3,601
Ethane	20,432	1,641	Toluene	17,620	4,284
Propane	19,944	2,385	Xylene	17,760	4,980
n-Butane	19,680	3,113	Methanol	9,078	768
Isobutane	19,629	3,105	Ethanol	11,929	1,451
n-Pentane	21,091	3,709	Ammonia	8,001	365
Isopentane	19,478	3,716	Hydrogen sulfide	6,545	596
n-Hexane	19,403	4,412			

Source: Adapted from Reynolds, J., Dupont, R., and Theodore, L., *Hazardous Waste Incineration Calculations*, John Wiley & Sons, New York, 1991.

5.7 VOC HEATING VALUE APPROXIMATION

There are more than 2 million organic compounds. Heats of combustion for some compounds are available in the literature. The heating value can be approximated from the chemical formula using the following equation:[5]

$$Hc = 15410 + 100 \times [((H \times 1.01/MW) \times 323.5 - (S \times 32.04/MW)$$
$$\times 115 - (O \times 32/MW) \times 200.1 - (Cl \times 35.45/MW) \qquad (5.1)$$
$$\times 162 - (N \times 14.01/MW) \times 120.5)]$$

where

Hc = heat of combustion (Btu/Lb – HHV)
H, S, O, Cl, N = number of atoms of H, S, O, Cl, N in the VOC compound
MW = molecular weight of the VOC compound

Example 5.4: A compound has the chemical formula $C_2H_4Cl_2$ (ethylene dichloride). Estimate its heating value.

H = 4
C = 2
Cl = 2
S = 0
O = 0
N = 0

Its molecular weight = 2×12.01 (C) $+ 4 \times 1.01$ (H) $+ 2 \times 35.45$ (Cl) $= 98.96$

$$Hc = 15,410 + 100 \times [((4 \times 1.01/98.96) \times 323.5 - (0 \times 32.04/98.96)$$
$$\times 115 - (0 \times 32/98.96) \times 200.1 - (2 \times 35.45/98.96) \times 162$$
$$- (0 \times 14.01/98.96) \times 120.5)]$$

$$Hc = 5,124 \text{ Btu/lb (HHV)}$$

The published value for the higher heating value of ethylene dichloride is 5398 Btu/lb.

A comparison of estimated and published heating values, using the method above for other compounds, is as follows:

Compound	Chemical Formula	Heat of Combustion (Btu/lb – HHV)	
		Published Value	Estimated
Napthalene	$C_{10}H_8$	17,298	17,449
Propane	C_3H_8	21,661	21,338
Isopropyl alcohol	C_3H_8O	14,776	9,105
Hydrogen cyanide	HCN	11,327	10,397
Pyridine	C_5H_5N	15,696	15,343
Methyl mercaptan	CH_4S	11,176	10,468

This method produces a good estimate, except for compounds that contain oxygen atoms. It is most accurate for hydrocarbons (compounds containing carbon and hydrogen only).

5.8 HEAT OF FORMATION

Another method that can be used to estimate heats of combustion when published data are not available is based on heats of formation. The heat of formation of any compound is defined as the enthalpy change during a chemical reaction where 1 mole of product is formed from its elements. It is also referenced to a standard temperature (usually 77°F). There are many engineering and technical books that contain heats of formation of organic compounds.

The heat of reaction is the difference between the sum of the heats of formation of the products of the reaction and the sum of the heats of formation of the reactants. As applied to thermal oxidation combustion calculations, the heat of reaction is the same as the heat of combustion. Thus,

$$\delta Hc = \Sigma H_f \text{ (products)} - \Sigma H_f \text{ (reactants)}$$

Let's use methane again as an example. The reaction chemistry is

$$CH_4 + 2 O_2 \rightarrow CO_2 + 2 H_2O$$

The published heats of formation for each compound are as follows:

$$\text{Methane } H_f = -32,200 \text{ Btu/lb-mol}$$

$$\text{Carbon dioxide } H_f = -169,294 \text{ Btu/lb-mol}$$

$$\text{Water vapor } H_f = -104,036 \text{ Btu/lb-mol (vapor state)}$$

$$\text{Oxygen } H_f = 0$$

Note that the heat of formation is zero for all pure elements (in this example oxygen). By normal convention, a negative sign indicates that heat is evolved (exothermic) when the compound is formed from its elements. Completing the calculation:

$$\Delta Hc = \Sigma H_f \{-169,294 + 2 \times (-104,036)\} - \Sigma H_f (-32,200)$$

$$\Delta Hc = -345,166 \text{ Btu/lb-mol}$$

Since the molecular weight of methane is 16.04 (lb/lb-mol), the heat of combustion is

$$Hc = 345,166/16.04 = 21,519 \text{ Btu/lb}$$

This is almost identical to the published value. Since the value of the heat of formation of water was for the vapor state, the heat of combustion is the LHV. If the liquid state heat of formation of water vapor were used, the heat of combustion would be the higher heating value. Unlike the previous methods that used the number of atoms of specific types to estimate the heat of combustion, this method produces theoretically exact results.

Combustion products almost always consist of the following compounds (shown with their heat of formation in units of Btu/lb-mol):

Compound	Heat of Formation (Btu/lb-mol)
CO_2	−169,294
H_2O	−104,036 (vapor)
SO_2	−127,728
N_2	0
HCl	−39,713

Example 5.5: Determine the heat of combustion of ethyl chloride. Its chemical formula is C_2H_5Cl and its heat of formation is − 48,240 Btu/lb-mol.

The oxidation reaction is

$$C_2H_5Cl + 3 O_2 \rightarrow 2 CO_2 + 2 H_2O + HCl$$

$$\Delta Hc = \Sigma H_f \{2 \times (-169,294) + 2 \times (-104,036) + (-39,713)\} - \Sigma H_f (-48,240)$$
$$(2 \ CO_2) \qquad\qquad (2 \ H_2O) \qquad\qquad (HCl) \qquad\qquad (C_2H_5Cl)$$

$$\Delta Hc = -538,133 \text{ Btu/lb-mol}$$

The molecular weight of ethyl chloride is 64.52. Therefore, the heat of combustion is

$$\Delta Hc = -554.173/64.52 = 8340 \text{ Btu/lb}$$

Again, this is the lower heating value, since the heat of formation used for water was for the vapor phase.

Now that oxidation chemistry and heat generation has been explained, the overall process will be described. It is best illustrated by referring to Figure 5.1. If the waste gas is at an elevated temperature (i.e., greater than the reference temperature), its heat content (enthalpy) is first determined. As described earlier, this is determined by multiplying its mass (lb/hr) by its heat capacity (Btu/lb-°F) and multiplying this product by the temperature differential (actual temperature vs. reference temperature).

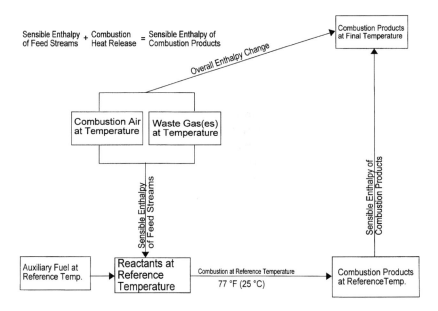

Figure 5.1   Mass and heat balance schematic.

Example 5.6: The coke oven gas of Example 5.2 is at an initial temperature of 300°F before injection into a thermal oxidizer. Its flow rate is 632 scfm. Determine its initial enthalpy.

Again, its composition is as follows:

Component	Vol%
Carbon monoxide	4.5
Hydrogen	57.9
Methane	30.3
Nitrogen	5.5
Carbon dioxide	1.8

The overall molecular weight of this gas is 9.62. A reference temperature of 77°F will be used. The mean heat capacities for each of these components are

Component	Btu/lb-mol-°F
Carbon monoxide	6.98
Hydrogen	6.91
Methane	9.29
Nitrogen	6.97
Carbon dioxide	9.26

The flow rate of 632 scfm is equivalent to 100 lb-mol/hr, determined as follows:

$$\frac{632 \text{ scf}}{\text{min}} \times \frac{60 \text{ min}}{\text{hr}} \times \frac{1 \text{ lb-mol}}{379 \text{ scf}} = 100 \text{ lb-mole/hr}$$

The initial enthalpy (heat content) of each component of the coke oven gas is then

Component	Quantity (lb-mol/hr)	Enthalpy (Btu/hr)
Carbon monoxide	4.5	$4.5 \times 6.98 \times (300 - 77) = 7{,}004$
Hydrogen	57.9	$57.9 \times 6.91 \times (300 - 77) = 89{,}220$
Methane	30.3	$30.3 \times 9.29 \times (300 - 77) = 62{,}772$
Nitrogen	5.5	$5.5 \times 6.97 \times (300 - 77) = 8{,}549$
Carbon dioxide	1.8	$1.8 \times 9.26 \times (300 - 77) = 3{,}717$
		Total = 171,262

Therefore, the heat content of the waste stream represented by the left vertical line in Figure 5.1 is 171, 262 Btu/hr in this example. Referring again to Figure 5.1, the organic (VOC) components of the waste gas are combusted at the reference temperature (77°F). From Example 5.2, the heat released was determined to be 450 Btu/scf when these gases were thermally oxidized. The 632 scfm of this example represents a total combustion heat release of

$$\frac{632 \text{ scf}}{\text{min}} \times \frac{60 \text{ min}}{\text{hr}} \times \frac{450 \text{ Btu}}{\text{scf}} = 17{,}064{,}000 \text{ Btu/hr}$$

This heat will not be released without the addition of oxygen for combustion. The oxygen required is determined by balancing the chemical oxidation reaction for each combustible component of the waste gas. For the coke oven gas, the combustible components are carbon monoxide (CO), hydrogen (H_2), and methane (CH_4). Their oxidation reactions are as follows:

	CO	+	$1/2 O_2 \rightarrow$	CO_2
lb-mol/hr	4.5		$1/2 \times 4.5$	4.5
lb/hr	126		72	198

	H_2	+	$1/2 O_2 \rightarrow$	H_2O
lb-mol/hr	57.9		$1/2 \times 57.9$	57.9
lb/hr	117		926	1043

	CH$_4$ +	2 O$_2$ →	CO$_2$ +	2 H$_2$O
lb-mol/hr	30.3	2 × 30.3	30.3	2 × 30.3
lb/hr	486	1939	1334	1091

The total oxygen added for the reactions is 91.8 lb-mol/hr or, in other units,

$$\frac{91.8 \text{ lb-mol}}{\text{hr}} \times \frac{32 \text{ lb O}_2}{\text{lb-mol}} = \frac{2939 \text{ lb}}{\text{hr}}$$

or

$$\frac{91.8 \text{ lb-mol}}{\text{hr}} \times \frac{379 \text{ scf}}{\text{lb-mol}} = \frac{34{,}792 \text{ scf}}{\text{hr}} = \frac{579.9 \text{ scf}}{\text{min}}$$

Typically, the oxygen supplied for the combustion reactions is provided in the form of air. Air is composed of approximately 79% nitrogen (by volume) and 21% oxygen (by volume). Therefore, to add 91.8 lb-mol/hr of oxygen requires the addition of 91.8/0.21 = 437.1 lb-mol/hr of air. This will include the addition of 0.79 × 437.1 = 345.3 lb-mol/hr of nitrogen.

These reaction products, along with the inert gases (N$_2$, CO$_2$) present in the coke oven gas, represent the combustion products at the reference temperature of 77°F. The overall combustion products at this stage of the process are then

Component	(lb-mol/hr)
Carbon dioxide	4.5 + 30.3 +1.8 = 36.6
Water vapor	57.9 + 60.6 = 118.5
Nitrogen	5.5 + 345.3 = 350.8

The last step shown in Figure 5.1 is heating the products of combustion to the preestablished thermal oxidizer operating temperature. This is illustrated by the following example:

Example 5.7: Raise the temperature of the coke oven gas combustion products from Example 5.6 to 2200°F.

This is done in two steps: (1) determine the heat required, and (2) add that heat, using auxiliary fuel if the heat release from the VOCs in the waste stream itself is inadequate to achieve the final temperature.

Step 1 determines the change in enthalpy needed.

Component	Quantity (lb-mol/hr)	Enthalpy (Btu/hr)
Carbon dioxide	36.6	36.6 × 12.12 × (2200 − 77) = 941,746
Water vapor	118.5	118.5 × 9.48 × (2200 − 77) = 2,384,936
Nitrogen	350.8	350.8 × 7.60 × (2200 − 77) = 5,660,088
		Total = 8,986,770

Note that the combustion air added does not directly enter into these equations. It is indirectly accounted for by the increase in the mass of the combustion products. The nitrogen component of the air is directly added to the combustion products. The oxygen component of air reacts and becomes part of the mass of combustion products. So while not shown directly, the addition of combustion air is taken into account in these calculations.

The sensible heat already contained in the coke oven gas is 171,262 Btu/hr as calculated previously. Combined with the combustion heat release of 17,064,000 Btu/hr, the total heat available (without the addition of auxiliary fuel) is 17,235,262 Btu/hr; but only 8,986,770 Btu/hr is needed. Therefore, even without the addition of any auxiliary fuel, there is 8,248,492 Btu/hr of excess heat.

Thus far, the combustion has taken place at stoichiometric conditions (exact amount of theoretical oxygen required). However, to ensure complete reaction, excess oxygen (air) is always added. This air produces an additional sensible heat load on the process and will tend to reduce the temperature. Usually, excess air is added to a thermal oxidation process to produce 3% (by volume) of oxygen in the combustion products. The compositions of the stoichiometric combustion products are

Component	Quantity (lb-mol/hr)	Vol%
Carbon dioxide	36.6	7.23
Water vapor	118.5	23.42
Nitrogen	350.8	69.35
Total =	505.9	100

Note that lb-mol/hr is a unit of volume. Thus volume percents can be calculated directly.

To obtain 3% oxygen by volume in the combustion products, air will be added. The quantity required is calculated as follows:

$$\text{Let } N = \text{air added}$$

$$0.03 = 0.21 \times N / (N + 505.9)$$

$$0.03\,N + 0.03 \times (505.9) = 0.21 \times N$$

$$0.03\,N + 15.177 = 0.21\,N$$

$$15.177 = (0.21\,N - 0.03\,N)$$

$$15.177 = 0.18\,N$$

$$84.32 \text{ lb-mol/hr} = N$$

Therefore, 84.32 lb-mol/hr of air must be added, which includes $0.21 \times 84.32 = 17.71$ lb-mol/hr of oxygen. The composition of the combustion products is now:

Component	Quantity (lb-mol/hr)	Vol%
Carbon dioxide	36.6	6.20
Water vapor	118.5	20.08
Nitrogen	350.8 + 66.61	70.72
Oxygen	17.71	3.00
Total =	590.22	100

The above equation can be simplified and made more generic as follows:

If

A = Combustion products at stoichiometric (lb-mol/hr)

B = Desired O_2 concentration in the final combustion products (vol%)

C = Air that must be added (lb-mol/hr)

Then

$$C = (A \times B/100)/(0.21 - B/100) \tag{5.2}$$

This assumes that air is the source of the oxygen added and that there is no oxygen in the combustion products before the addition of this air.

With the addition of the 84.32 lb-mol/hr ($84.32 \times 28.85 = 2433$ lb/hr) of air, the excess heat previously calculated will be reduced. The heat required to raise the temperature of the 84.32 lb-mol/hr of air is

$$Q = 84.32 \text{ lb-mol/hr} \times 7.69 \text{ Btu/lb-mol}°F \times (2200°F - 77°F)$$

$$Q = 1,376,597$$

where 7.69 Btu/lb-mol-°F is the mean heat capacity for air between 77 and 2200°F (Table 5.1).

Since this value is less than the 8,248,942 Btu/hr of excess heat calculated before the addition of the air, the actual temperature will rise above 2200°F, even with the addition of the excess air. As will be discussed in later chapters, allowing the temperature to rise too high requires the use of more expensive refractory lining in the thermal oxidizer and thus increases its cost. It is usually more cost-effective to quench the combustion products.

The problem is solved by setting the final temperature at 2200°F and determining the amount of air that must be added to reduce the temperature to 2200°F. Air added to reduce temperature is called "quench air." The quench air added must be equivalent to the excess heat. This balance in equation form is

$$Q_{excess} = Q_{quench}$$

$$8{,}248{,}942 \text{ Btu/hr} = m_a \times Cp_{air} \times (T_f - 77)$$

where

    m_a = mass of air (lb-mol/hr)

    Cp_{air} = heat capacity of air (Btu/lb-mol-°F)

    $8{,}248{,}942 \text{ Btu/hr} = m_a \times 7.69 \text{ Btu/lb-mol-°F} \times (2200 - 77)$

    $8{,}248{,}942 \text{ Btu/hr} = 16{,}325.87 \times m_a$

    $m_a = 505.23 \text{ lb-mol/hr} = 14{,}577 \text{ lb/hr} = 3191 \text{ scfm}$

While this air satisfies the heat balance to maintain the temperature at 2200°F, it exceeds the amount previously calculated to increase the oxygen content of the combustion products to 3%. However, in general, this additional air is not detrimental as long as it is not injected in a manner that quenches the combustion reactions before they can proceed to completion. Thus, it is best to add quench air after the combustion zone.

The discussion to this point has focused on adiabatic (no heat loss) calculations. In real systems, some heat is lost to the shell of the thermal oxidizer, and this must be taken into account in the energy balance. The concept of heat loss through the shell and its calculation is discussed in Chapter 7.

5.9   WATER QUENCH

Water can also be used for quenching combustion products when the heat released by the VOCs themselves exceeds the heat required. A second approach to the problem above would be to add the air required to achieve 3% oxygen in the combustion products, followed by quench water to control the temperature.

It was shown previously that 84.32 lb-mol/hr of air was needed to raise the oxygen concentration of the coke oven gas stoichiometric combustion products to 3% O_2 (by volume). The addition of this air produces a heat load of 1,376,597 Btu/hr. This reduces the overall excess heat to 6,872,345 Btu/hr. Water quench not only produces a sensible heat load but also a latent heat load for evaporation of the liquid water. For this case,

$$\text{Excess heat (Btu/hr)} = m \times (Cpl \times (212 - 60) + Hv + Cpg \times (T_f - 212))$$

where m   = mass of water injected (lb/hr)

    Cpl = heat capacity of liquid water (1.0 Btu/lb-°F)

    Hv   = heat of vaporization of water (970.5 Btu/lb-°F at 212°F)

    Cpg = mean heat capacity of water vapor between 212°F and the final temperature (Btu/lb-°F)

    T_f = final temperature (°F)

An initial water temperature of 60°F was assumed in the calculations above. Although the heat capacity of liquid water varies with temperature, a value of 1.0 Btu/lb-°F can be used with very little error. Substituting the numbers for this example into the equation produces the following:

$$6{,}872{,}345 \text{ Btu/hr} = m \times (1.0 \times (212 - 60) + 970.5$$
$$+ (9.48/18.02) \times (2200 - 212))$$

$$m = 3169 \text{ lb/hr or } 6.34 \text{ gal/min}$$

where 9.48 Btu/lb-mol-°F is the mean heat capacity of water vapor between 212 and 2200°F.

Thus, 3169 lb/hr of water plus 84.32 lb-mol/hr (2433 lb/hr) of air is equivalent to 505.23 lb-mol/hr (14,577 lb/hr) of air for quenching these combustion products.

While water quench does require the addition of more equipment to a thermal oxidizer, it has the advantage of lowering the volume of combustion products and thus the volume of the residence chamber needed to attain a specified residence time. In the above examples, using air quench alone produced 6387 scfm of combustion products (at 2200°F), while the combination of air and water quench produced combustion products totaling 4839 scfm (at 2200°F). Thus, the volume of the residence chamber required is reduced by approximately 25% with the water quench.

5.10 AUXILIARY FUEL ADDITION

In the coke oven gas example, no additional fuel was needed to raise the coke oven gas to the specified operating temperature. In VOC applications, more often than not, additional fuel is required. The mass and energy balance calculations must not only account for the heat energy supplied by this fuel but also for the heat load imposed by any combustion air added with it. The following example illustrates a case where auxiliary fuel is needed.

Example 5.8: A waste gas at 77°F consists of 1% acetone and 99% nitrogen. Its flow rate is 100 scfm. The acetone will be destroyed in a thermal oxidizer operating at 1500°F. Determine the amount of auxiliary fuel required. Natural gas with a heating value of 950 Btu/scf (LHV) will be used as the auxiliary fuel.

Since the waste gas is at 77°F, its initial enthalpy (referenced to 77°F) is 0. The only combustible component of the waste gas is acetone. Its heating value is 12,593 Btu/lb (LHV). The heat released by oxidizing the acetone at the reference temperature is

Acetone chemical formula = C_3H_6O
Acetone MW = $3 \times 12.01 + 6 \times 1.01 + 16.00 = 58.09$
100 scfm × 0.01 = 1 scfm of acetone
1 scfm × 60 min/hr × 1 lb-mol/379 scf × 58.09 lb/lb-mol = 9.2 lb/hr
Combustion heat release = 12,593 Btu/lb × 9.2 lb/hr = 115,856 Btu/hr

The POCs are

	C_3H_6O +	$4\,O_2$ →	$3\,CO_2$ +	$3\,H_2O$
Scfm	1.0	4.0	3.0	3.0

lb-mol/hr	0.158	0.633	0.474	0.474
lb/hr	9.2	20.2	20.9	8.5

The combustion products will also contain 99 scfm (15.67 lb-mol/hr) of nitrogen that was originally present in the waste gas plus the nitrogen in the air needed to supply the oxygen required for combustion. Air contains 79% nitrogen by volume and 21% oxygen, or 3.76 volumes of nitrogen per volume of oxygen. The nitrogen added in this case is, therefore, 0.633 lb-mol/hr × 3.76 = 2.38 lb-mol/hr. Therefore, the total nitrogen in the combustion products is 15.67 + 2.38 = 18.05 lb-mol/hr. Again, to convert from scfm to lb-mol/hr, multiply the scfm times 60 (convert scfm to scfh) and divide by 379 (1 lb-mol/379 scf).

To raise these products of combustion to 1500°F requires additional heat, calculated as follows:

Component	Quantity (lb-mol/hr)	Enthalpy (Btu/hr)
Carbon dioxide	0.474	$0.474 \times 11.41 \times (1500 - 77) = 7,696$
Water vapor	0.474	$0.474 \times 9.93 \times (1500 - 77) = 6,697$
Nitrogen	(15.67 + 2.38) = 18.05	$18.05 \times 7.35 \times (1500 - 77) = 163,893$
		Total = 203,180

Since the heat released (115,856 Btu/hr) is less than the heat required (203,180 Btu/hr), auxiliary fuel must be added. The net heat required is 203,180 − 115,856 = 87,324 Btu/hr.

It may seem that the additional heat required can be determined simply by dividing the net heat required by the heating value of natural gas or {(87,324 Btu/hr)/(950 Btu/scf)}. A cursory evaluation indicates that the product of this division is the natural gas required in units of scf/hr. However, since the waste gas contains no oxygen, oxygen must be added in the form of air to oxidize the natural gas. This produces an additional heat load that must be overcome by the addition of still more natural gas, and the cycle continues.

This problem can be approached in one of several ways. One is to calculate the natural gas required, including its stoichiometric quantity of air. This is an iterative procedure since the air included with the natural gas provides an additional heat load which must be accounted for by the addition of more natural gas to overcome that heat load, and so on. However, after a number of iterations, the heat balance can be closed.

Another method is to account for the air added when calculating the amount of natural gas required. The formula that must be solved when using this method is as follows:

$$\text{Natural gas (lb-mol/hr)} = \text{heat required (Btu/hr)}/\{LHV(Btu/scf)/\text{lb-mol}$$
$$- (Cp_{CO_2} + 2 \times Cp_{H_2O} + 7.52 \times Cp_{N_2}) \times (T - 77)\}$$

where

LHV = lower heating value of the natural gas (Btu/scf)

Cp_{CO_2}, Cp_{H_2O}, Cp_{N_2} = heat capacities of carbon dioxide, water vapor, and nitrogen, respectively (Btu/lb-mol-°F)

This equation is not as complex as it may first appear. The denominator takes into account the fact that products of combustion of the natural gas will contain carbon dioxide (1 lb-mol/lb-mol natural gas), water vapor (2 lb-mol/lb-mol natural gas), and nitrogen from the nitrogen in the combustion air (7.52 lb-mol/lb-mol natural gas). The fact that these constituents must also be heated to the thermal oxidizer operating temperature reduces the amount of heat available from the natural gas to raise the waste stream to this temperature.

Continuing with the example:

Natural gas (lb-mol/hr) = 87,324 Btu/hr/{950 Btu/scf × 379 scf/lb-mol
   − [(11.41 + 2 × 9.96 + 7.52 × 7.35) × (1500 − 77)]}

$$\text{Natural gas required (lb-mol/hr)} = \frac{87,324 \text{ Btu/hr}}{(360,050 - 123,235)}$$

$$= 0.369 = 140 \text{ scfh}$$

Therefore, the total heat input required is 140 scfh × 950 Btu/scf = 132,858 Btu/hr. This result indicates that the thermal oxidation process needs 132,858 Btu/hr of natural gas, despite the 62,430 Btu/hr heat released from oxidation of the waste gas itself.

The previous calculations illustrate the concept of available heat. Available heat is the actual amount of the total heat input that can be applied to the process. It can be specified in absolute terms (Btu/scf) or relative terms (percent of heat input). It is a function of the fuel used and the corresponding excess air rate. Tables 5.4 to 5.6 show values of available heats for natural gas at various excess air levels and heating values.

TABLE 5.4
Available Heat — Natural Gas (900 Btu/scf)

Temp (F)	Excess Air (%)	Available Heat (%)	Available Heat (Btu/scf)
600	0	87.46	787
600	10	86.38	777
600	25	84.77	763
600	50	82.08	739
600	100	76.71	690
600	200	65.96	594
800	0	82.53	743
800	10	81.04	729
800	25	78.81	709
800	50	75.09	676
800	100	67.65	609
800	200	52.76	475

TABLE 5.4 (CONTINUED)
Available Heat — Natural Gas (900 Btu/scf)

Temp (F)	Excess Air (%)	Available Heat (%)	Available Heat (Btu/scf)
1000	0	77.47	697
1000	10	75.56	680
1000	25	72.69	654
1000	50	67.91	611
1000	100	58.35	525
1000	200	39.24	353
1200	0	72.28	650
1200	10	69.93	629
1200	25	66.42	598
1200	50	60.56	545
1200	100	48.85	440
1200	200	25.42	229
1400	0	66.97	603
1400	10	64.19	578
1400	25	60.01	540
1400	50	53.06	478
1400	100	39.15	352
1400	200	11.33	102
1600	0	61.55	554
1600	10	58.33	525
1600	25	53.48	481
1600	50	45.41	409
1600	100	29.27	263
1800	0	56.04	504
1800	10	52.36	471
1800	25	46.84	422
1800	50	37.64	339
1800	100	19.24	173
2000	0	50.43	454
2000	10	46.29	417
2000	25	40.09	361
2000	50	29.76	268
2000	100	9.08	82
2200	0	44.73	403
2200	10	40.14	361
2200	25	33.25	299
2200	50	21.78	196

TABLE 5.5
Available Heat — Natural Gas (950 Btu/scf LHV)

Temp (F)	Excess Air (%)	Available Heat (%)	Available Heat (Btu/scf)
600	0	88.12	837
600	10	87.10	827
600	25	85.57	813
600	50	83.03	789
600	100	77.93	740
600	200	67.75	644
800	0	83.45	793
800	10	82.04	779
800	25	79.93	759
800	50	76.40	726
800	100	69.35	659
800	200	55.25	525
1000	0	78.65	747
1000	10	76.84	730
1000	25	74.13	704
1000	50	69.60	661
1000	100	60.55	575
1000	200	42.44	403
1200	0	73.74	700
1200	10	71.52	679
1200	25	68.19	648
1200	50	62.64	595
1200	100	51.54	490
1200	200	29.35	279
1400	0	68.71	653
1400	10	66.07	628
1400	25	62.12	590
1400	50	55.53	528
1400	100	42.35	402
1400	200	16.00	152
1600	0	63.58	604
1600	10	60.52	575
1600	25	55.93	531
1600	50	48.29	459
1600	100	32.99	313
1600	200	2.41	23

TABLE 5.5 (CONTINUED)
Available Heat — Natural Gas (950 Btu/scf LHV)

Temp (F)	Excess Air (%)	Available Heat (%)	Available Heat (Btu/scf)
1800	0	58.35	554
1800	10	54.87	521
1800	25	49.64	472
1800	50	40.92	389
1800	100	23.49	223
2000	0	53.04	504
2000	10	49.12	467
2000	25	43.24	411
2000	50	33.45	318
2000	100	13.87	132
2200	0	47.64	453
2200	10	43.29	411
2200	25	36.77	349
2200	50	25.90	246
2200	100	4.15	39

TABLE 5.6
Available Heat — Natural Gas (1000 Btu/scf)

Temp (F)	Excess Air (%)	Available Heat (%)	Available Heat (Btu/scf)
600	0	88.71	887
600	10	87.74	877
600	25	86.29	863
600	50	83.87	839
600	100	79.04	790
600	200	69.36	694
800	0	84.28	843
800	10	82.94	829
800	25	80.93	809
800	50	77.58	776
800	100	70.88	709
800	200	57.48	575
1000	0	79.72	797
1000	10	78.00	780

continued

TABLE 5.6 (CONTINUED)
Available Heat — Natural Gas (1000 Btu/scf)

Temp (F)	Excess Air (%)	Available Heat (%)	Available Heat (Btu/scf)
1000	25	75.42	754
1000	50	71.12	711
1000	100	62.52	625
1000	200	45.32	453
1200	0	75.05	750
1200	10	72.94	729
1200	25	69.78	698
1200	50	64.51	645
1200	100	53.96	540
1200	200	32.88	329
1400	0	70.27	703
1400	10	67.77	678
1400	25	64.01	640
1400	50	57.75	578
1400	100	45.23	452
1400	200	20.20	202
1600	0	65.40	654
1600	10	62.49	625
1600	25	58.13	581
1600	50	50.87	509
1600	100	36.34	363
1600	200	7.29	73
1800	0	60.43	604
1800	10	57.12	571
1800	25	52.15	522
1800	50	43.88	439
1800	100	27.32	273
2000	0	55.38	554
2000	10	51.66	517
2000	25	46.08	461
2000	50	36.78	368
2000	100	18.18	182
2200	0	50.25	503
2200	10	46.12	461
2200	25	39.93	399
2200	50	29.60	296
2200	100	8.95	89

Continuing with the previous example, with the natural gas added, the waste stream temperature would be raised to the 1500°F target temperature. However, there is no excess oxygen in the combustion products. Typically, thermal oxidizers are operated with at least 3% of oxygen (by volume) in the combustion products. This oxygen is normally added in the form of excess air. The amount required can be calculated into two ways. The first is to add an arbitrary quantity of air, add natural gas to raise this air to the target temperature, add additional air for the natural gas, and continue this iterative process until the desired oxygen concentration is obtained, while satisfying the heat balance. This procedure is illustrated on the logic diagram of Figure 5.2. While this is the most accurate method of performing the mass and energy balance, it is tedious and is best performed with a computer program or spreadsheet.

The second method of performing the heat balance is to use the concept of available heat. By firing the natural gas with excess air, the oxygen in the excess supplies the oxygen needed to achieve the desired oxygen level in the combustion products. However, this is also a trial-and-error method, since the oxygen concentration produced by a certain level of excess air is unknown.

Referring to Table 5.5, natural gas with a LHV of 950 Btu/scf will be used to complete the previous example. In that example, 87,324 Btu/hr of heat was needed to raise the waste gas to its final temperature of 1500°F. If 50% excess

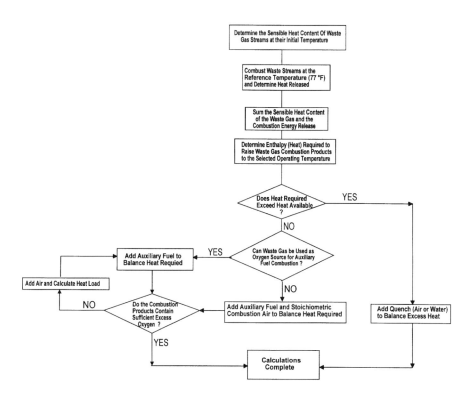

Figure 5.2   Logic diagram for mass and energy balance calculations.

air is used as a first trial, interpolating between 1400 and 1600°F provides an available heat value of 494 Btu/scf of heat available. Therefore, the quantity of natural gas needed is

$$\text{Natural gas (scfh)} = \{(87{,}324 \text{ Btu/hr})/(494 \text{ Btu/scf})\} = 176.8$$

This is equivalent to a total heat input from natural gas of 167,960 Btu/hr. At stoichiometric conditions, natural gas requires 9.52 scf air/scf of natural gas. The total air added is $176.8 \times 9.52 \times 1.5 = 2525$ scfh. The excess air is $0.5 \times 9.52 \times 176.8 = 842$ scfh. Of this amount, 21% is oxygen or 177 scfh. To summarize, combustion of this quantity of natural gas produces

	CH_4 +	$2\,O_2 \rightarrow$	CO_2 +	$2\,H_2O$
Scfh	177	2×177	177	2×177
lb-mol/hr	0.467	2×0.467	0.467	2×0.467

In addition, the unreacted (excess) air produces 0.79×2525 (none reacted) = 1995 scfh (5.26 lb-mol/hr) of nitrogen and $842 \times 0.21 = 177$ scfh (0.467 lb-mol/hr) of oxygen. The totals for the waste stream combustion products plus the natural gas combustion products are as follows:

Component	Quantity (lb-mol/hr)	Vol%
Carbon dioxide	0.474 +0.467	3.96
Water vapor	0.474 + 0.934	5.93
Nitrogen	15.67 +5.26	88.14
Oxygen	0.466	1.96
Total =	23.75	100

As shown, 50% excess air was not sufficient to attain 3% oxygen in the combustion products. Repeating the process with 100% excess air produces the following results:

Component	Quantity (lb-mol/hr)	Vol%
Carbon dioxide	0.474 +0.467	3.62
Water vapor	0.474 + 0.934	5.42
Nitrogen	15.67 +7.02	87.37
Oxygen	0.933	3.59
Total =	25.97	100

The natural gas rate is 244 scfh. A slightly lower excess air rate would achieve the desired result of 3.0% O_2.

5.11 ADIABATIC FLAME TEMPERATURE

The concept of excess air has been alluded to several times in this chapter. Again, excess air is needed to ensure complete (or nearly complete) VOC oxidation and

TABLE 5.7
Natural Gas Theoretical Flame Temperatures and Combustion Products (LHV = 950 Btu/scf)

Excess Air (%)	Theoretical Flame (°F)	Combustion Products (vol%)			
		CO_2	H_2O	N_2	O_2
15	3525	8.34	16.68	72.41	2.56
20	3425	8.04	16.08	72.65	3.23
25	3325	7.75	15.49	72.88	3.88
30	3230	7.47	14.94	73.10	4.49
35	3139	7.21	14.42	73.30	5.06
40	3053	6.97	13.94	73.49	5.59
45	2973	6.75	13.50	73.67	6.08
50	2895	6.54	13.07	73.84	6.55
60	2753	6.15	12.31	74.14	7.40
75	2565	5.66	11.31	74.53	8.50
100	2305	4.99	9.97	75.06	9.98
150	1922	4.03	8.06	75.82	12.10
200	1654	3.38	6.76	76.33	13.53

low CO emissions. In a well-designed thermal oxidation system, the combustion products exiting the thermal oxidizer should be homogeneous. However, in the upstream section of the oxidizer where waste gas, air, and auxiliary fuel are added, localized reactions occur which produce heterogeneous concentration profiles of these reactants. Thus, to ensure complete VOC destruction, excess oxygen (usually in the form of air) is added to ensure that each VOC and auxiliary fuel molecule comes in contact with an oxygen molecule.

In the U.S., most thermal oxidizer applications use natural gas as the primary fuel. Although its composition can vary over a limited range as discussed in Chapter 4, Table 5.7 presents adiabatic flame temperatures and combustion products for a typical natural gas at various excess air rates. Adiabatic flame temperature is defined as the highest possible temperature a combustion reaction can attain, assuming no heat losses and no dissociation. The products of combustion of natural gas are carbon dioxide and water vapor. At temperatures above 3000°F, these reaction products begin to dissociate into carbon monoxide and hydrogen, adsorbing heat in the process. At 3500°F, about 10% of the CO_2 in a typical flame dissociates to CO and O_2. Heat absorption consumes 4347 Btu/lb of CO formed and 61,100 Btu/lb of hydrogen formed. However, as the gas cools, the dissociated CO and H_2 recombine with O_2 and liberate the heat adsorbed during dissociation. Therefore, the heat is not lost. However, the overall effect is to lower the actual flame temperature. The degree with which this dissociation occurs at various temperatures is shown in Table 5.8. The theoretical flame temperature for other fuels and organic compounds with stoichiometric air addition is shown in Table 5.9. The theoretical flame temperatures for fuel oil at various excess air rates are presented in Table 5.10.

TABLE 5.8
Dissociation of Combustion Products

Reaction	Temp (°F)	% Dissociation
$O_2 \leftrightarrow 2\ O^a$	4937	5.95
$N_2 \leftrightarrow 2\ N^a$	7142	5.0
$H_2O \leftrightarrow H_2 + 1/2\ O_2$	3182	0.37
	3992	4.1
$CO_2 \leftrightarrow CO + 1/2\ O_2$	2048	0.014
	2804	0.400
	3993	13.5

[a] Indicates unstable radical species.

Source: *Combustion and Incineration Processes*, Niessen, W.R., Marcel Dekker, New York, 1995.

TABLE 5.9
Theoretical Flame Temperatures (Stoichiometric Air)

Combustible Gas	Theoretical Flame Temp (°F)
Carbon monoxide	4311
Hydrogen	3960
Methane	3640
Ethane	3710
Propane	3770
Butane	3780
Pentane	3720
Hexane	3710
Ethylene	3910
Propylene	3830
Acetylene	4250
Benzene (vapor)	3860
Hydrogen cyanide	4250
Methanol (vapor)	3610
Ammonia	3440
Refinery gas (1365 Btu/scf – LHV)	3841
Refinery gas (591 Btu/scf – LHV)	3832

5.12   EXCESS AIR

All thermal oxidizers use a burner for producing a high-temperature flame. The burner mixes auxiliary fuel with combustion air to produce this flame. The burner

TABLE 5.10

No. 2 Fuel Oil Theoretical Flame Temperatures and Combustion Products

Excess Air (%)	Theoretical Flame (°F)	Combustion Products (vol%)			
		CO_2	H_2O	N_2	O_2
15	3468	11.96	10.11	75.00	2.92
20	3361	11.49	9.71	75.16	3.62
25	3260	11.06	9.35	75.31	4.27
30	3165	10.66	9.01	75.44	4.88
40	2989	9.94	8.40	75.68	5.96
50	2832	9.32	7.87	75.89	6.91
60	2691	8.76	7.41	76.07	7.75
75	2505	8.05	6.80	76.31	8.83
100	2248	7.08	5.98	76.64	10.29
150	1873	5.71	4.83	77.09	12.37
200	1610	4.78	4.04	77.40	13.77

is operated continuously for modulating the temperature of the thermal oxidizer or, in some cases, is only needed for start-up. Burner design and operation will be described in more detail in Chapter 7.

Typically, a burner is operated with excess air. However, some burners can be operated with as little as 50% of the stoichiometric air required. This substoichiometric operation is sometimes used when the waste gas itself contains a significant concentration of oxygen. The oxygen in the waste stream provides the oxygen source to completely oxidize the substoichiometric combustion products from the burner. This concept will be discussed further in later chapters, particularly with regard to control of NOx emissions. Table 5.11 shows the theoretical temperature for substoichiometric combustion of natural gas at various combustion air levels. The quantity

TABLE 5.11

Theoretical Flame Temperatures (Substoichiometric Combustion of Natural Gas)

Stoichiometric Air (%)	Theoretical Temp (°F)
50	2369
61	2757
70	3065
80	3315
91	3510
100	3640

TABLE 5.12
Combustion Constants for Common Organic Gases

Compound	Chemical Formula	Molecular Weight	Heat of Combustion Btu/scf High	Low	Btu/Lb High	Low	Stoichiometric Air (Mol/mol organic) (scfm/scfm organic)
Hydrogen	H_2	2.02	325	275	61,100	51,623	2.38
Carbon monoxide	CO	28.01	322	322	4,347	4,347	2.38
Methane	CH_4	16.04	1,013	913	23,879	21,520	9.53
Ethane	C_2H_6	30.07	1,792	1,641	22,320	20,412	16.68
Propane	C_3H_8	44.09	2,590	2,385	21,661	19,944	23.82
n-Butane	C_4H_{10}	58.19	3,370	3,113	21,308	19,680	30.97
Isobutane	C_4H_{10}	58.19	3,363	3,105	21,257	19,629	30.97
n-Pentane	C_5H_{12}	72.14	4,016	3,709	21,091	19,517	38.11
Isopentane	C_5H_{12}	72.14	4,008	3,716	21,052	19,478	38.11
n-Hexane	C_6H_{14}	86.17	4,762	4,412	20,940	19,403	45.26
Ethylene	C_2H_4	28.05	1,614	1,513	21,644	20,295	14.29
Propylene	C_3H_6	42.08	2,336	2,186	21,041	19,691	21.44
n-Butene	C_4H_8	56.10	3,084	2,885	20,840	19,496	28.59
Isobutene	C_4H_8	56.10	3,068	2,869	20,730	19,382	28.59
n-Pentene	C_5H_{10}	70.13	3,836	3,586	20,712	19,363	35.73
Benzene	C_6H_6	78.11	3,751	3,601	18,210	17,480	35.73
Toluene	C_7H_8	92.13	4,484	4,284	18,440	17,620	42.88
Xylene	C_8H_{10}	106.16	5,230	4,980	18,650	17,760	50.02
Methanol	CH_4O	32.04	868	768	10,259	9,078	7.15
Ammonia	NH_3	17.03	441	363	9,668	8,001	3.57
Hydrogen sulfide	H_2S	34.08	647	596	7,100	6,545	7.15

of air required for theoretical combustion varies by fuel type and atomic composition. Combustion constants for a variety of fuels and common organic compounds are shown in Table 5.12.

The heating value of a gas mixture, either higher or lower, can be derived from its components using this table. The volume fraction of each component is multiplied by its heating value to obtain the overall heating value of the gas mixture. This is illustrated in the following example.

Example 5.9: Natural gas has the following component composition: methane 94.06%, nitrogen 0.28%, carbon dioxide 0.36%, ethane 3.92%, n-butane 0.20%, isobutane 0.12%, n-pentane 0.09%, isopentane 0.10%, and propylene 0.87%. Determine its higher and lower heating values in units of Btu/scf.

The lower and higher heating values of each of these components are as follows:

Component	LHV (Btu/scf)	HHV (Btu/scf)
Methane	913	1013
Ethane	1641	1792
n-Butane	3113	3370
Isobutane	3105	3363
n-Pentane	3709	4016
Isopentane	3716	4008
Propylene	2186	2336

The lower heating value is

$$\text{LHV (Btu/scf)} = 0.9406 \times 913 + 0.0392 \times 1641 + 0.002 \times 3113$$
$$+ 0.0012 \times 3105 + 0.0009 \times 3709 + 0.001 \times 3716 + 0.0087 \times 2186$$
$$\text{LHV (Btu/scf)} = 959$$

The higher heating value is

$$\text{HHV (Btu/scf)} = 0.9406 \times 1013 + 0.0392 \times 1792 + 0.002 \times 3370$$
$$+ 0.0012 \times 3363 + 0.0009 \times 4016 + 0.001 \times 4008 + 0.0087 \times 2336$$
$$\text{HHV (Btu/scf)} = 1062$$

Even though the atomic structure, stoichiometric air required, and heating value vary widely for fuels, organic compounds, and VOCs, the amount of air required per million Btu of heat release is similar with few exceptions. This is illustrated in Table 5.13. With the exception of hydrogen and carbon monoxide, the quantity of (stoichiometric) air required (scfm) per million Btu of heat release falls in the range of 155 to 174. Thus, as a rough check of detailed combustion calculations, the

TABLE 5.13
Combustion Air Requirements

Gas	LHV (Btu/scf)	scf of gas/MM Btu released	Stoichiometric Air Requirements	
			scf of air/scf of gas	scfm air/MM Btu/hr heat
Hydrogen	275	3636	2.38	144
Carbon monoxide	322	3106	2.38	123
Methane	913	1095	9.53	174
Ethane	1641	609	16.68	169
Propane	2385	419	23.82	166
Butane	3113	321	30.97	166
Pentane	3709	270	38.11	171
Hexane	4412	227	45.26	171
Ethylene	1513	661	14.29	157
Propylene	2186	457	21.44	163
Benzene	3601	278	35.73	165
Toluene	4284	233	42.88	167
Methanol	768	1302	7.15	155
Ethanol	1451	689	14.29	164
Ammonia	365	2740	3.57	163

amount of stoichiometric air required should be approximately 160 scfm/MM Btu of heat release if hydrogen or carbon monoxide are not present. (Note: MM is typically used in the combustion industry to denote "million.")

5.13   WET VS. DRY COMBUSTION PRODUCTS

Many times the products of combustion are reported on a dry basis. This is common because many analytical instruments/procedures used to measure the composition of the combustion products first extract/condense the water vapor before performing the measurement. The following formulas can be used to convert from wet basis to dry basis and vice versa.

POC component N (dry) = POC component N (wet) × $1/(1 - (H_2O/100))$

POC component N (wet) = POC component N (dry) × $(1 - (H_2O/100))$

where H_2O = actual water vapor concentration (vol%)

Wet and dry basis oxygen concentrations for natural gas combustion at various excess air levels are shown in Table 5.14. This table also illustrates the fact that for a given fuel, the ratio of combustion air to gas combusted (vol/vol) is equivalent to a specific oxygen concentration in the combustion products. The measurement of oxygen concentration in the combustion products is important since typically a specific quantity (%) is desired. However, analytical instruments are available to measure these combustion products on both a wet or dry basis. It is important to understand which type of instrument is being used for a particular application.

TABLE 5.14
Comparison of Wet and Dry Oxygen
Concentrations for Natural Gas Combustion

Air/Fuel Ratio	% Excess Air	% O_2 in POC (Wet)	% O_2 in POC (Dry)
10	4.9	0.90	1.10
11	15.4	2.57	3.09
12	25.9	3.99	4.72
13	36.4	5.21	6.07
14	46.9	6.26	7.22
15	57.4	7.18	8.21
16	67.9	7.99	9.06
17	78.4	8.72	9.80
18	88.9	9.36	10.46
19	99.4	9.94	11.05
20	109.9	10.47	11.57

5.14   SIMPLIFIED CALCULATIONAL PROCEDURES

While relatively exact calculational procedures have been described to this point, there are approximate methods that, although not as accurate, are adequate in many cases. One equation for estimating temperature, in various forms, is[4]

$$T = 60 + (LHVo)/[0.3 \times \{1 + (1 + EA) \times 7.5 \times 10^{-4} \times LHVo\}\ ]$$

$$EA = \frac{[(LHVo/\{0.3 \times (T - 60)\})] - 1}{(7.5 \times 10^{-4}) \times (LHVo)}$$

where

T = temperature (°F)
EA = excess air fraction (% excess air/100)
$LHVo$= lower heating value (Btu/lb)

Example 5.10: What is the temperature resulting from natural gas combustion at 100% excess air. Its lower heating value is 950 Btu/scf or 22,392 Btu/lb. The initial temperatures of the natural gas and combustion air are both 60°F.

Substituting into the equation above

$$T = 60 + (22,392)/[0.3 \times \{1 + (1 + (100/100)) \times 7.5 \times 10^{-4} \times 22,392\}]$$

$$T = 2218°F$$

This compares to the value of 2305°F given in Table 5.7. A limitation of this equation is that the initial temperature of the waste gas and combustion air must both be 60°F.

For the most part, even though they contain VOCs, waste gas(es) have predominantly inert components such as nitrogen, oxygen, carbon dioxide, and water vapor. The heat capacities of nitrogen, oxygen, and carbon dioxide are similar on a mass basis. For example, the mean heat capacities of these gases between 77 and 1500°F are 0.262, 0.242, and 0.259 Btu/lb-°F, respectively. The mean heat capacity of water vapor is 0.496 Btu/lb-°F in this temperature range. Unless the waste gas is an off-gas from a scrubber where it is saturated with water vapor, the water content of most waste gases is less than 10%. Therefore, a mean heat capacity of 0.28 Btu/lb-°F can be used as an approximation for most gaseous waste streams. The mean heat capacity of air is approximately 0.26 Btu/lb-°F in this same temperature range. Natural gas (essentially pure methane) has a mean heat capacity 0.82 Btu/lb-°F in this same temperature range. Since the mass rate of methane is usually small compared to the air and waste streams, its contribution to the heat load is sometimes ignored.

The heat required to raise a waste stream plus its combustion air to the specified final temperature is

$$Q = m_{wg} \times Cp_{wg} \times (T_f - T_I) + m_{air} \times Cp_{air} \times (T_f - T_I)$$

where

Q = heat required (Btu/lb)

m_{wg} = mass of waste gas (lb/hr)

Cp_{wg} = mean heat capacity of waste gas between T_f and T_I

T_f = final temperature (°F)

T_I = initial temperature (°F)

m_{air} = mass of air added (lb/hr)

Cp_{air} = mean heat capacity of air between T_f and T_I

But as an approximation, we have assumed that $Cp_{wg} = 0.28$ Btu/lb-°F and $Cp_{air} = 0.26$ Btu/lb-°F. Therefore,

$$Q = m_{wg} \times 0.28 \times (T_f - T_I) + m_{air} \times 0.26 \times (T_f - T_I)$$

If both the waste gas and combustion air are at ambient temperature, the equation can be further simplified to

$$Q = (m_{wg} \times 0.28 + m_{air} \times 0.26) \times (T_f - 77)$$

Previously, it was shown that the mass of air required could be approximated by 160 scfm/MM Btu of heat release. The above equation provides the heat requirement for stoichiometric combustion only. After this equation is solved, the concept of available heat can be applied. An excess air level is assumed, and the natural gas required is determined by dividing Q by the available heat. This is best illustrated by repeating Example 5.8.

Example 5.11: A waste gas at 77°F consists of 1% acetone and 99% nitrogen. Its flow rate is 100 scfm. The acetone will be destroyed in a thermal oxidizer operating at 1500°F. Determine the amount of auxiliary fuel required. Natural gas with a heating value of 950 Btu/scf (LHV) is used as the auxiliary fuel.

Acetone — C_3H_6O MW = $3 \times 12.01 + 6 \times 1.02 + 16 = 58.2$

Acetone mass rate = 1% /100 × 100 scfm × 60 min/hr × 1 lb-mol/379 scf ×
58.2 MW = 9.2 lb/hr

N_2 mass rate = 99% /100 × 100 scfm × 60 min/hr × 1 lb-mol/379 scf × 28.01
MW = 439 lb/hr

Acetone LHV = 12,593 Btu/lb

Acetone heat release = 12,593 Btu/lb × 9.2 lb/hr = 115,856 Btu/hr = 0.116
MM Btu/hr

Therefore, stoichiometric air estimate = 160 scfm × 0.115 MM Btu/hr = 18.4 scfm

18.4 scfm × 60 min/hr/379 scf/lb-mol × 28.85 (MW) = 84 lb/hr

Alternately, the mass rate of air can be determined as follows:

$$18.4 \text{ scfm} \times 60 \text{ min/hr} \times 0.075 \text{ lb/ft}^3 = 84 \text{ lb/hr}$$

(*Note*: The density of air at 77°F and 1 atmosphere pressure is 0.075 lb/ft³.) Substituting these values into the stoichiometric heat equation

$$Q = [\{(9.2 + 439) \times 0.28 + 84 \times 0.26\} \times (1500 - 77)]$$

$$Q = 209,659$$

However, the acetone itself releases 115,856 Btu/hr when oxidized. Therefore, the net amount still required is 209,659 – 115,856 = 93,803 Btu/hr. To complete the calculations, use an available heat equivalent to 50% excess air (494 Btu/scf).

$$Q \text{ (natural gas)} = (93,803 \text{ Btu/hr})/(494 \text{ Btu/scf})$$

$$Q = 207 \text{ scfh}$$

At 50% excess air, the excess oxygen is

Excess O_2 = 50%/100 × 9.52 × 21%/100 O_2 = 1 scf O_2/scf nat gas
Excess O_2 = 1 scf O_2/scf nat gas × 207 scfh nat gas = 207 scfh O_2
Nitrogen added = 1.5 × 9.52 × 0.79 = 11.28 scf/scf nat gas
Nitrogen added = 11.28 scf/scf nat gas × 207 scfh nat gas = 2335 scfh

Unfortunately, although excess oxygen is available, the concentration is not obvious. Fortunately, for natural gas combustion, not only does mass balance on both sides of the equation, but so does volume (this is not generally true for other fuels). Therefore, since the VOC concentration is dilute and does not have a significant impact on the mass balance, we can simply divide the oxygen excess by the total volume of each component.

Waste gas = 100 scfm = 6000 scfh
Natural gas = 207 scfh
Excess oxygen = 207 scfh
Nitrogen added = 2335 scfh

$$\% \ O_2 \text{ in POC} = \frac{207}{(6000 + 207 + 207 + 2335)} \times 100 = 2.37 \ \%$$

This equation can be generalized when natural gas is used as the auxiliary fuel as follows:

$$\% \ O_2 \text{ in POC} = (EX/100 \times 2 \times NG)/((WG + (\{EX/100 \times 2) + (1 + EX/100) \times 7.52\} \times NG) + NG)) \times 100$$

where

> WG = waste gas flow rate (scfh)
> EX = excess air rate (%)
> NG = natural gas rate (scfh)

If a specific oxygen concentration in the combustion products is required, this method must be repeated on a trial-and-error basis.

The calculations are much simpler if the waste gas is a VOC-contaminated air stream. Usually no additional air is needed and the issue of the amount of air that must be added is then avoided. The oxygen in the air stream provides the oxygen needed for both the VOC and auxiliary fuel combustion. In this case, the heat balance equation simplifies to

$$Q = m_{wg} \times Cp_{wg} \times (T_f - T_I)$$

Example 5.12: A waste gas at 77°F consists of 1% acetone in an air stream. The waste gas flow rate is 100 scfm. The acetone will be destroyed in a thermal oxidizer operating at 1500°F. Determine the amount of auxiliary fuel required. Natural gas with a heating value of 950 Btu/scf (LHV) is used as the auxiliary fuel. (Same as Example 5.8 except that acetone is in air stream instead of a nitrogen stream.)

Using values calculated from Example 5.8:

Acetone – C_3H_6O MW = 3 × 12.01 + 6 × 1.02 + 16 = 58.2
Acetone mass rate = 9.2 lb/hr
Acetone LHV = 12,593 Btu/lb
Acetone heat release = 12,593 Btu/lb × 9.2 lb/hr = 115,856 Btu/hr = 0.116
    MM Btu/hr

The mass of air is

> Air mass rate = 99% /100 × 100 scfm × 60 min/hr × 1 lb-mol/379 scf
>   × 28.85 MW = 452 lb/hr

or

$$99\%/100 \times 100 \text{ scfm} \times 60 \text{ min/hr} \times 0.076 \text{ lb/ft}^3 = 452 \text{ lb/hr}$$

$$Q = m_{wg} \times Cp_{wg} \times (T_f - T_I)$$

$$Q = 452 \text{ lb/hr} \times 0.28 \text{ Btu/lb-°F} \times (1500°F - 77°F)$$

$$Q = 180,094 \text{ Btu/hr}$$

Since the heat release from oxidation of the acetone in the waste stream is 115,856 Btu/Hr, the heat that must be supplied by auxiliary fuel is 180,094 – 115,856

= 64, 238 Btu/hr. This compares to 93,803 Btu/hr in Example 5.11. Since the oxygen needed for both VOC and auxiliary fuel combustion is already contained in the waste gas, no further calculations are needed.

If the waste stream and/or combustion air are already at an elevated temperature, the auxiliary fuel requirements are reduced. Sometimes elevated waste stream temperatures are a consequence of the process generating the waste stream. Other times, the waste stream is preheated through heat exchange with the combustion products (this is discussed in more detail in Chapter 8). The following example illustrates the effect of a higher initial waste stream temperature.

Example 5.13: A waste stream at 585°F consists of 1% acetone in an air stream. The waste stream flow rate is 100 scfm. The acetone will be destroyed in a thermal oxidizer operating at 1500°F. Determine the amount of auxiliary fuel required. Natural gas with a heating value of 950 Btu/scf (LHV) is used as the auxiliary fuel (same as Example 5.12 except that the initial waste gas temperature is elevated).

$Q = m_{wg} \times Cp_{wg} \times (T_f - T_I)$
$Q = 452$ lb/hr $\times 0.28$ Btu/Lb-°F $\times (1500°F - 585°F)$
$Q = 115,802$ Btu/hr

Since the heat release from oxidation of the acetone in the waste gas is 115,856 Btu/hr, no additional heat from auxiliary fuel is required to raise the temperature of the waste stream to 1500°F.

These simpler methods are usually adequate for obtaining approximate mass and energy balances for the purpose of generating budgetary cost estimates or to establish very preliminary equipment sizes. They will generally produce results within ±15% of the true values. They can be performed with a hand calculator while the more rigorous procedure described by Figure 5.2 requires the use of a computer program.

6   Waste Characterization and Classification

CONTENTS

Thermal oxidation technology can be used to treat VOC-contaminated waste streams that vary over a wide range of flow and composition. As a consequence, most thermal oxidation systems are custom designed for a given application. The composition of the waste stream has a very significant impact on the final design. This chapter will discuss the constituents normally encountered in industrial waste streams and how these constituents affect the ultimate design of the thermal oxidizer.

6.1   WASTE STREAM CHARACTERIZATION

There are two ways to characterize a waste stream: direct and indirect. In the direct method, each specific compound in the waste stream is identified along with its concentration. An indirect method is to provide an ultimate analysis of the waste stream. An ultimate analysis describes a waste in terms of its individual atoms plus ash. It is typically specified as carbon, hydrogen, oxygen, nitrogen, sulfur, chlorine, and ash. These constituents are usually listed on a weight percent basis.

Example 6.1: The compound nitromethane has the chemical formula of CH_3NO_2. Its molecular weight is 61.05. Determine its ultimate analysis.

The weight fraction (ultimate analysis) of each atom in its structure is

Atom	Weight (%)
Carbon	$1 \times 12.01/61.05 = 19.67$
Hydrogen	$3 \times 1.01/61.05 = 4.96$
Nitrogen	$1 \times 14.01/61.05 = 22.96$
Oxygen	$2 \times 16.0/61.05 = 52.42$

Ultimate analyses are routinely used with solid waste which typically are heterogeneous mixtures of multiple species. Determining the ultimate analysis of such mixtures is usually much easier than determining the individual components. However, the ultimate analysis has several drawbacks. First, unless specified on a dry basis, it is not clear whether any hydrogen and oxygen present are from water or are part of the chemical structure. Water vapor is inert (no chemical reaction or heat release), while hydrogen and oxygen are normally part of the oxidation chemistry. The same is true for hydrogen chloride (HCl) and sulfur dioxide (SO_2). It is not possible to determine from the ultimate analysis whether these are part of the waste stream in the inert (noncombusting) forms of HCl or SO_2 or individual atoms that participate in the combustion reactions. Also, while heats of combustion can be estimated based on an ultimate analysis, they will not be as exact as tabulated values for individual organic species.

If the only information available is the ultimate analysis of an organic waste stream, it can be converted to a "pseudo" chemical compound by dividing the weight fractions of each element by their respective molecular weights and normalizing to the atom with the lowest mole weight.

Example 6.2: The ultimate analysis of an organic waste is 58.80 wt% carbon, 9.87 wt% hydrogen, and 31.33 wt% oxygen. Convert to a "pseudo" chemical compound.

Atom	Wt%	Atomic Weight	Mole Fraction
Carbon	58.80	12.01	$58.80/12.01 = 4.90$
Hydrogen	9.87	1.01	$9.87/1.01 = 9.79$
Oxygen	31.33	16.0	$31.33/16.0 = 1.96$

It would be perfectly correct to describe the chemical formula as $C_{4.9}H_{9.79}O_{1.96}$. However, most scientists and engineers prefer to work with whole numbers. Dividing each atom by the atom with the lowest mole number (1.96) results in the following formula: $C_{2.5}H_5O$. To eliminate the fractional quantity of carbon, multiply each atom fraction by 2, which results in the following formula: $C_5H_{10}O_2$. This formula could represent isopropyl acetate. However, this illustrates another problem with the use of ultimate analysis: different compounds can have the same molecular structure. These are called isomers. For example, vinyl alcohol and acetaldehyde both have the chemical formula C_2H_4O but have different molecular structures.

6.2 WASTE STREAM VARIABILITY

Many times, thermal oxidation equipment vendors are presented with a waste stream composition at a single flow rate and single composition. Very few processes produce waste streams that do not have some variability. The single flow and composition often represent an average for both. In fact, while averages may provide a good representation of fuel requirements under normal conditions, they do not provide the equipment designer with the most crucial information, the variability of the waste flow and composition. The equipment must be designed for the extremes, not the averages. For example, start-up can represent the condition requiring the highest auxiliary fuel requirement by virtue of the fact that the waste stream is not yet contaminated with VOCs and thus provides no heat of its own. Many equipment vendors design the auxiliary fuel burner for the maximum waste gas mass flow assuming no heat contribution from VOCs present. Conversely, spikes in VOC concentration can increase the overall heat release and elevate the temperature. Air or water quench may be required under that scenario. Therefore, it is important to note not just normal, average, or typical flows and compositions, but also their variability when specifying equipment or analyzing a process as a potential application for thermal oxidation technology.

6.3 MINOR CONTAMINANTS — MAJOR PROBLEMS

6.3.1 ORGANIC COMPONENTS

It is important to identify all of the constituents of a waste stream, even if they are present in minor quantities. Minor quantities of certain components can have a larger impact on the design and operation of a thermal oxidation system than inert components in much higher concentrations.

The halogen group of atoms has been discussed frequently in prior chapters. Thermal oxidation of halogenated organic compounds produces acid gases such as hydrogen chloride. When present in combustion gases, they can condense where local temperatures are at or below their dewpoint and be very corrosive to metals. Some can even attack the refractory lining of a thermal oxidizer. These species can troublesome in concentrations of less than 25 ppmv. Sulfur and phosphorus oxide species in the waste stream can have the same affect as the halogens. The emission of halogenated acids and sulfur dioxide is also limited by environmental regulations. If these are present in large quantities, a gas scrubber may be required at the exit of the thermal oxidizer.

6.3.2 INORGANIC COMPONENTS

Inorganic (noncombustible) components of a waste gas can also have a major impact on the design of a thermal oxidizer. This can be due to both chemical and physical processes. Chemically, substances such as sodium can react with the refractory lining, causing it to melt, spall, or disintegrate. These inorganic species either enter the thermal oxidizer as a particulate or are converted to solid oxides during the combustion process. The postcombustion form of inorganic trace metals is shown in Table 6.1.

TABLE 6.1
Postcombustion Oxidation State of Trace Metals

Inorganic Metal	Element Symbol	Postcombustion Form
Lead	Pb	PbO/Pb_3O_4
Chromium	Cr	Cr_2O_3
Cadmium	Cd	CdO
Beryllium	Be	BeO
Arsenic	As	As_2O_5
Copper	Cu	CuO/Cu_2O
Manganese	Mn	Mn_3O_4/Mn_2O_3
Iron	Fe	Fe_2O_3/Fe_3O_4
Nickel	Ni	NiO
Tin	Sn	SnO
Zinc	Zn	ZnO
Cobalt	Co	CoO/Co_3O_4
Vanadium	V	V_2O_5
Mercury	Hg	Hg

Some of the more common inorganic species that can chemically attack a refractory are sodium, potassium, and vanadium compounds. These materials form complex salts with silica and other refractory components, resulting in the formation of low melting compounds (fluxing).

Physical attack can occur through erosion or abrasion. Here, solid particles impact the surface of the refractory at high velocity over a prolonged period of time, causing the refractory to wear away. In the high temperature environment of a thermal oxidizer, particulate can become molten. Molten particulate can adhere to the surface of a thermal oxidizer or downstream equipment, buildup over time, and eventually constrict the flow of combustion products. If the melting temperature of the particulate is known (for example, of a single component inorganic particulate) then design features can be included in the thermal oxidizer design to prevent this build-up.

One such design approach is to quench the combustion products to below the melting point of the particulate. An example is shown in Figure 6.1. Here, a waste

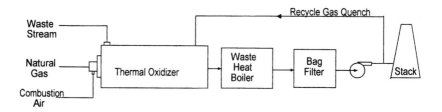

Figure 6.1   Quench of combustion products to solidify molten particulate.

heat boiler is installed to recover heat from the thermal oxidizer. To prevent molten particulate from adhering to boiler tubes, cooled recycle gas from the exit of the boiler is mixed with the combustion products to solidify the particulate before it enters the boiler. Melting points of some inorganic species typically encountered in thermal oxidizer applications are shown in Table 6.2.

TABLE 6.2
Melting Points of Selected Inorganic Compounds and Eutectics

Compound	Chemical Formula	Melting Point (°F)
Calcium hydroxide	$Ca(OH)_2$	1075
Calcium sulfate	$CaSO_4$	2640
Calcium chloride	$CaCl_2$	1440
Potassium chloride	KCl	1420
Potassium sulfate	K_2SO_4	1960
Potassium hydroxide	KOH	680
Sodium hydroxide	$NaOH$	600
Sodium chloride	$NaCl$	1475
Sodium sulfate	Na_2SO_4	1625
Sodium carbonate	Na_2CO_3	1564

The problem can become more complex if more than one inorganic species is present. Sometimes pure inorganic components mix to form a lower melting point mixture. These compounds are called **eutectics**. Melting points of common eutectic forms are shown in Table 6.3. Adding silicon and aluminum oxides from the refractory lining to the mix can lower these temperatures even further. The presence of phosphorus and iron can also exacerbate the problem.

TABLE 6.3
Melting Points of Selected Eutectics

Eutectic Mixture	Melting Point (°F)
$MgSO_4/Na_2SO_4$	1220
$Na_2O/SiO_2/Na_2SO_4$	1175
$NaCl/Na_2SO_4$	1153
$NaCl/Na_2CO_3$	1171
Na_2SO_4/Na_2CO_3	1522
$NaCl/Na_2SO_4/Na_2CO_3$	1134

The point of this discussion is that while these minor species usually have an insignificant effect on the mass and energy balance for a thermal oxidizer, they can have a large impact on both the thermal oxidizer and downstream emissions control systems. Therefore, it is important to identify these minor species along with the major components of the waste gas.

6.4 CLASSIFICATIONS

Most gaseous waste stream can be grouped into one of three main classifications:

1. Contaminated air streams
2. Contaminated inert gas streams
3. Rich gas streams

The process design strategy for each of these gaseous waste classifications follows.

6.4.1 CONTAMINATED AIR STREAMS

This first category of gaseous waste streams encompasses air streams contaminated with low concentrations of volatile organic compounds. Smoke and odors are typical examples of this type of waste gas stream. These gas streams are characterized by a minimum oxygen concentration of 18% (volume) and no possibility of an explosion hazard (the organic compound concentration is well below the lower flammable limit (less than 25% of the LEL — discussed in detail in Chapter 14). Gases with lower oxygen levels, higher organic compound concentrations, and/or corrosive compounds and particulates fall into one of the other two classifications.

Examples of industries and processes that generate contaminated air streams are

Coating operations
Printing lines
Paint booths
Textile converters
Textile finishing
Solvent cleaning
Packaging
Food processing and baking
Pulp and paper
Computer chip manufacturing (semiconductor)
Drying
Ventilation odors
Certain chemical processes
Rendering plants
Sewage treatment plants

The thermal oxidizer process design uses the contaminated air stream as the combustion air source for the burner. For applications with

- low flow rates (less than 5000 scfm)
- no corrosive compounds present
- a clean gas stream (no particulates)
- gaseous auxiliary fuel such as natural gas, propane, or liquefied petroleum gas (LPG)

the most economical (equipment cost) design option uses an in-duct burner (also known as a raw gas burner, duct burner, or in-line burner — discussed further in Chapter 7). As mentioned above, the contaminated air stream is used as the combustion air source. It passes through the burner mixing plates and is thoroughly mixed with the fuel and ignited as shown in Figure 6.2.

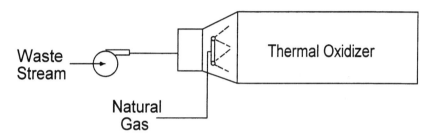

Figure 6.2   Direct injection of waste stream through in-line (duct) burner.

The resulting products of combustion are raised to the required operating temperature and held at that temperature for a time sufficient for the VOC thermal oxidation reactions to near completion. The burner design and a "profile plate" create the necessary turbulence. This design is called "direct flame incineration." The contaminants pass directly through a flame front. It is a highly effective form of destruction for volatile organic compounds (VOCs). It is also thermally efficient in that no outside air is introduced into the system. The typical requirements and features for this type of process (system) include:

- A contaminated air stream without any sulfur compounds, halogenated compounds, and/or particulates
- A contaminant concentration less than 25% of the lower explosive limit (LEL)
- An oxygen concentration greater than or equal to 18% (volume) at ambient temperature
- Gaseous fuels only (natural gas, propane, LPG, etc.)
- A waste gas pressure at the burner between 0.5 and 3 in. w.c. (a blower can provide the motive force if necessary)
- A maximum inlet gas temperature of 1000°F (this temperature is limited by the burner materials of construction)

- A maximum operating temperature (POC outlet temperature) of 1700°F; this temperature is also limited by the burner materials of construction
- A maximum fuel turndown requirement of 20:1
- A maximum contaminated air flow turndown of 3:1

With a cold waste gas stream, the minimum oxygen concentration required to use this type of burner is approximately 18%. However, if the temperature is above ambient (by recuperative heat recovery, for example), the oxygen concentration can be as low as 12% (if, for example, the waste stream temperature is 1000°F). The effect of temperature on the minimum oxygen concentration required is shown in Figure 6.3 for one manufacturer's duct (line) burner. While preheating the waste gas is feasible, the maximum operating temperature of this burner is 1700°F, due to heat radiation back to the metal structure of the burner. The refractory baffle, nozzle mix burner described in Chapter 7 does not suffer from this limitation but is limited by the volume of waste gas that can be injected through it.

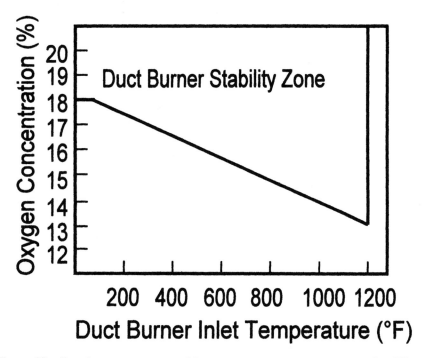

Figure 6.3   Duct burner waste stream inlet temperature vs. oxygen concentration (%).

When liquid fuels (#2 fuel oil, Bunker C, kerosene, waste liquids, etc.) or dual fuels are used, or when the oxygen content of the contaminated air stream is between 16 and 18% (volume) at ambient temperature, a nozzle mix burner must be used in lieu of the duct burner. Similar to the duct burner, the nozzle mix burner can be operated at lower oxygen concentrations if the temperature is elevated. Low oxygen

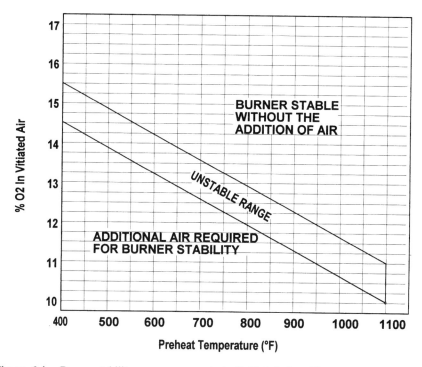

Figure 6.4 Burner stability vs. oxygen content of vitiated air oxidant.

concentration combustion air streams are termed "vitiated air." Figure 6.4 shows the stable operating range for a generic burner.

Depending upon the contaminated air flow rate and the burner design, all or part of the air stream is used as combustion air and passes through the burner. In most instances, the entire contaminated air stream cannot be introduced through the burner. Low flame temperatures, unstable combustion, high carbon monoxide (CO) emissions, high levels of products of incomplete combustion (PICs), and nuisance shutdowns would occur.

For most nozzle mix burner applications, the air stream should be split into a primary air stream and a secondary air stream. The primary air stream is introduced into the burner as the combustion air. The secondary air stream enters the system downstream of the burner and acts as a quench air or dilution air. This is illustrated in Figure 6.5.

The thermal efficiency of this system is also very high. In fact, the efficiency is equal to that of the in-duct burner system discussed above. No outside air is added to the system. However, only a portion of the contaminated air stream, the primary air stream, passes through the flame front (direct flame incineration). The secondary air stream does not enter the burner. It is injected downstream of the flame front and must be properly mixed with the burner products of combustion. Suitable turbulence is necessary to oxidize the contaminants in the secondary air stream.

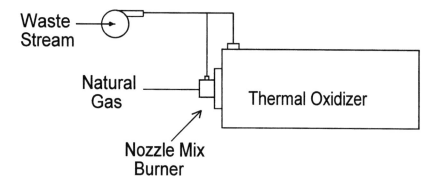

Figure 6.5   Waste stream split between nozzle mix burner and thermal oxidizer.

Achieving proper mixing and turbulence is not always straightforward. Injecting or adding the secondary air stream into the burner flame can quench the flame too quickly. This can lead to high CO emissions, high levels of products of incomplete combustion, low destruction efficiencies, and unstable combustion. A number of designs and configurations are acceptable for introducing secondary air into the system. One of the best methods is to "jet" the secondary air into the burner products of combustion well downstream of the flame root. Multiple jets or nozzles with sufficient velocity to penetrate the hot burner POCs is the preferred approach. The process requirements for this system are as follows:

- The contaminated air stream does not include any particulates, condensable gases, corrosive compounds, and/or acids
- The contaminant concentration is less than 25% of the LEL
- The oxygen concentration is greater than or equal to 16% (volume) at ambient conditions
- Liquid and/or gaseous auxiliary fuels
- The pressure of the contaminated air stream at the burner (inlet) must be between 4 and 8 in. w.c. (A blower normally provides the motive force.)
- The maximum inlet gas temperature is only limited by the burner materials of construction; an inlet temperature of 649°C (1200°F) or higher is not unusual
- The maximum operating temperature (POC) is only limited by the combustion chamber materials of construction; a maximum operating temperature of 1204°C (2200°F) is typical
- The maximum fuel turndown is 20:1 for gaseous fuel and 6:1 for liquid fuels
- The maximum contaminated air flow turndown is 5:1; this turndown is usually limited by the blower design and mixing requirements

When the contaminated air stream contains particulates, condensable gases, corrosive compounds, and/or acids, when little or no air pressure drop can be allowed, or when the air temperature is higher than the burner materials of construction can withstand, a nozzle mix burner is used with ambient (outside) combustion air. The contaminated air is introduced downstream of the burner and mixed with the burner products of combustion. In order to conserve fuel, the burner should be operated with very low excess air, or a specially designed substoichiometric air/fuel burner should be supplied. As mentioned above, mixing (turbulence) is the key to maximizing the destruction of the contaminants, reducing CO emissions, minimizing PICs, and maintaining reliable operation. The optimum design depends upon the fume composition, fume temperature, and turndown requirements. The most economical and straightforward design is to mount the burner radially or at an angle to the contaminated air flow. The burner then fires across the contaminated air stream and the air stream passes through the burner flame. A schematic of this arrangement is shown in Figure 6.6.

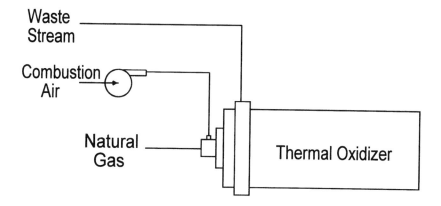

Figure 6.6   Waste system injected downstream of auxiliary fuel burner.

This design achieves destruction efficiencies of over 95%. With this design, the contaminated air stream can have inlet temperatures up to 1204°C (2200°F) or higher, and operating temperatures up to 1204°C (2200°F) or higher. These maximum temperatures are determined by the desired destruction level, materials of construction, upstream process conditions, blower design, and/or heat recovery requirements. "Jetting" the contaminated air stream into the burner POCs can result in higher destruction efficiencies. However, the air stream must be under a sufficient positive pressure and free of particulates. Sticky particulates are of most concern. They can eliminate the jetting design from consideration.

Example: 6.3: A 5000-scfm air stream at ambient temperature is contaminated with 100 ppmv of methyl chloride and also traces of hydrogen chloride. The thermal oxidizer should be designed to operate at a temperature of 1600°F to achieve high methyl chloride destruction efficiency. Natural gas with a lower heating value of 942 Btu/scf is used as the auxiliary fuel. The burner is operated at 25% excess air. Determine the process configuration.

Because the air stream contains hydrogen chloride gas, a duct (line) burner cannot be used. A nozzle mix burner will be used and operated at the specified 25% excess air level, using part of the waste stream to supply this air. The waste stream is split between the burner and thermal oxidizer inlet plenum, as shown in Figure 6.4. Using the methods outlined in Chapter 5, the auxiliary fuel required for this example is 170 scfm or 9.6 MM Btu/hr. Since natural gas requires 9.52 scfm of air per scfm of natural gas, at 25% excess, the air (from the waste stream) required in the burner is $170 \times 9.52 \times 1.25 = 2023$ scfm.

Example 6.4: The waste stream of Example 6.3 is found to contain particulate. Determine the process configuration.

In this case, the waste stream cannot be injected through either of the two types of burners (duct or nozzle mix). To minimize fuel consumption, a nozzle mix burner is fired at 50% of the stoichiometric air required for complete combustion of the natural gas. The natural gas firing rate required is 203 scfm, or 11.5 MM Btu/hr. In comparison to Example 6.3, the auxiliary fuel firing rate has increased by 1.9 MM Btu/hr or 20% higher. This is due to the increased heat load of the ambient air used in the burner. If the burner were fired at 25% excess with ambient air (instead of substoichiometric), the total firing rate required would be 16.4 MM Btu/hr.

While there are applications for both duct burners and nozzle mix burners, the cost of the duct burner is generally less for a given firing rate. However, since they are exposed directly to the gas stream and from the radiant heat of the flame, their applications are more limited than the nozzle mix burner. Burner design will be discussed further in Chapter 7.

6.4.2   CONTAMINATED INERT GAS STREAMS

Waste steams in this classification have less than 8% (volume) oxygen with low concentrations of organic compounds. The gas heating value is generally less than 30 Btu/scf. Since the oxygen level is below 8%, flashback and explosions are not a concern. The contaminant concentration and the oxygen concentration are not sufficient to develop and sustain a flame front (see Chapter 14). The waste stream is essentially inert.

"Inert" in the context of this discussion refers to a nonflammable mixture. Ignition of a VOC cannot occur if the oxygen concentration of the mixture is too low. The oxygen concentration threshold depends on the nature of the remaining inert gases and the particular VOC or organic compound and its temperature. The maximum permissible oxygen concentration to prevent ignition of flammable mixtures at ambient temperature is shown in Table 6.4 for selected VOCs.

The explosive potential of a waste stream is a function of the VOC concentration compared to its LEL. Preventing premature ignition of the VOC is a primary consideration in the design of a thermal oxidizer. Almost all applications contain provisions to prevent the VOC concentration from exceeding 50% of the LEL and many prevent this concentration from reaching 25% of the LEL.

TABLE 6.4
Maximum Safe Oxygen Concentrations to Prevent Ignition (Ambient Temperature)

Compound	Nitrogen Diluent	Carbon Dioxide Diluent
Acetaldehyde	12	
Acetone	13.5	15.5
Ammonia	15	
Benzene	11	14
Butadiene	10	13
Butane	12	14.5
Butene	11.5	14
Carbon disulfide	5	8
Carbon monoxide	5.5	6
Cyclopropane	11.5	14
Dimethylbutane	12	14.5
Diethyl ether	10.5	13
Ethane	11	13.5
Ethanol	10.5	13
Ethyl acetate	11.2	
Ethylene	10	11.5
Gasoline	11.5	14
Hexane	12	14.5
Hydrogen	5	6
Hydrogen sulfide	7.5	11.5
Isobutane	12	15
Kerosene	11	14
Methane	12	14.5
Methanol	10	13.5
Methyl ethyl ketone	11.4	
Methyl chloride	15	
n-Heptane	11.5	14
Pentane	11.5	14.5
Propane	11.5	14
Propylene	11.5	14
Toluene	9.1	
Xylene	8	

Source: Compiled from *National Fire Protection Association Bulletin* 69-1986 and *Chemical Engineering,* June 1994.

For example, a waste stream that is predominantly nitrogen and is contaminated with acetaldehyde will not ignite if the oxygen concentration is below 12%. Except for carbon monoxide and hydrogen, most organic compounds have a minimum oxygen concentration for ignition of greater than 8%. Safety considerations in the design of a thermal oxidizer are described in Chapter 14.

Typical examples of industries and processes that generate this type of waste gas stream include:

Pulp and paper
Inert drying or curing operations
Scrubber off-gases
Ceramic industry
Man-made fiber manufacturing
Absorber off-gases
Resin manufacturing (interim state)
Diesel/engine exhaust
Heat treating furnaces
Asphalt manufacturing
Chemical industry
Petrochemical industry

The process design for this type of waste gas stream uses a conventional burner firing a liquid or gaseous fuel. Ambient air is used for combustion air. The contaminated inert gas stream is introduced into the combustion chamber downstream of the burner flame root. Figure 6.6 illustrates this design concept.

Depending upon the inert gas characteristics (composition, flow, and temperature), the gas can enter the combustion chamber through a single tangential nozzle, a single radial nozzle, a single axial nozzle, or through a series of nozzles. Relatively clean gas streams, free of particulates, sticky tars, oil mists, and/or smokes can be introduced through multiple nozzles. The inert gas is "jetted" into the burner products of combustion. This jetting consists of injecting the waste gas through nozzles at a high velocity so that the gases intermix with the flame-front from the burner.

The recommended method for introducing a dirty gas with particulates, sticky tars, oil mists, and/or smokes is a single nozzle. If the fume contains between 3 and 8% (volume) oxygen, the auxiliary fuel burner can be operated substoichiometrically to conserve fuel. The inert gas stream provides the balance of the oxygen required for complete combustion.

Note that the ambient air blower must be sized for the maximum burner firing rate plus sufficient excess ambient air to maintain a minimum of 3% (volume, wet) excess oxygen in the POCs for all operating cases. This includes the start-up, warm-up, and shutdown conditions. The process requirements for this system are as follows:

- The inert gas stream does not include any particulates or condensable gases
- The contaminant concentration is low; the heating value of the gas is less than 30 Btu/scf
- The oxygen concentration is less than 8% (volume)
- Liquid and/or gaseous fuels are used as auxiliary fuel
- The pressure of the contaminated inert gas stream at the inlet nozzle must be between 4 and 8 in. w.c. (A blower can provide the motive force.)

The maximum inlet gas temperature is only limited by the inlet nozzle(s) materials of construction. The maximum operating temperature (POC outlet temperature) is only limited by the combustion chamber materials of construction. A maximum operating temperature of 2200°F is typical. The maximum fuel turndown is 20:1 for gaseous fuel and 6:1 for liquid fuels. The inert gas flow turndown is very high and can be zero with certain designs.

Example 6.5: A 25,000-scfm waste gas stream contains 2.5% carbon monoxide, 0.5% propane, 2.2% carbon dioxide, 6.5% water vapor, 80.3% nitrogen, and 8% oxygen. Its temperature is 250°F. The thermal oxidizer operating temperature is set at 1600°F. Determine its design configuration.

A quick determination of the heating value of the waste gas from its two combustible components produces 20 Btu/scf (0.025 × 322 Btu/scf + 0.005 × 2385 Btu/scf). Since its oxygen content is relatively low, the waste stream cannot be used as combustion air for the burner. To conserve fuel, the burner is operated at 50% of the stoichiometric air required. The oxygen content of the waste stream is then used to complete combustion. Using methods described in Chapter 5, 294 scfm of auxiliary fuel is required plus 1400 scfm of ambient combustion air for operation of the burner at 50% of stoichiometric air. The process configuration and rates are shown in Figure 6.7.

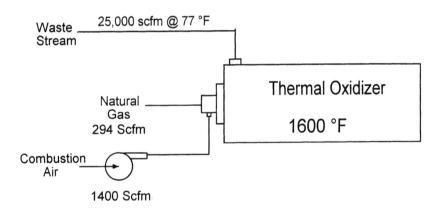

Figure 6.7   Example 6.5 flow distribution.

6.4.3   RICH GAS STREAMS

Streams in this classification have a very low oxygen content, a high percentage of combustible (VOC) compounds, and a heating value greater than 50 Btu/scf. Since the oxygen concentration is very low, flashback is not a concern. The oxygen concentration is not sufficient to develop and sustain a flame front. Any energy released prior to the combustion zone is not large enough to create an explosion.

Typical examples of industries and processes that generate this type of waste stream include:

Pulp and paper
Tank vents
Scrubber off-gases
CO gas
Reactor exhaust gases
Process upset gases
Resin kettle off-gases
Blast furnace gases
Heat treating furnaces
Stripper off-gases
Chemical industry
Petrochemical industry
Landfill gas

The design approach for this type of stream is to use the waste gas as a fuel in the burner. Usually, a specially designed low-Btu gas burner is used. In certain cases, a dual or multifuel burner is the appropriate selection. The waste gas is ducted to the burner and "fired" through a central nozzle or injector assembly. Combustion air is introduced around the waste nozzle. The burner design, especially the waste injector assembly, is critical to achieving proper and stable combustion and nuisance-free operation. The burner design is also crucial to achieving high levels of VOC destruction.

In most of the applications involving rich gas streams, the composition and flow rate of the waste gas can change significantly over time. During process start-up, shutdown, or upset conditions, the waste stream may become "inert" or the flow may decrease below the design turndown limit. Unstable combustion can occur or the flame can be extinguished. In order to prevent this from occurring, a secondary auxiliary burner is provided. This configuration is depicted in Figure 6.8.

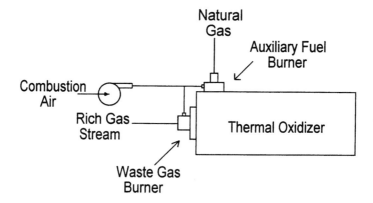

Figure 6.8   Rich waste stream process configuration.

The auxiliary burner uses a conventional fuel and fires during all operating conditions. It provides a positive, stable, and robust ignition source. It also maintains the necessary operating temperature when the "rich" waste stream becomes "inert" or when the waste flow is shut off or decreases below the design turndown limitations.

Rich waste streams typically can sustain combustion without (or with minimal) auxiliary fuel. If the waste gas heating value (LHV) is greater than 150 Btu/scf and does not contain high concentrations of water vapor, this is certainly true. Waste gases between 50 and 150 Btu/scf may or may not be able to sustain combustion without auxiliary fuel addition. A minimal amount of auxiliary fuel should be used to establish a strong ignition point if the heating value is less than 200 Btu/scf. The waste heat content then provides the necessary energy for the oxidation reactions to proceed. In fact, with this type of waste, the minimum operating temperature is not an issue. The heat released from the waste tends to drive the temperature upward. The maximum temperature is really the concern with these gases and is typically set at 2200°F to minimize refractory costs. Excess air through the burner is used to control this temperature.

Example 6.6: A 350-scfm waste stream at ambient temperature has the following composition: 12.8% pentane, 21.6% benzene, 1.8% toluene, 24.8% butane, and 39% nitrogen (all by volume). Determine the design and operating conditions for a thermal oxidizer.

The heating value of this gas mixture is 2102 Btu/scf determined as follows:

Cumbustible Component	Heating Value (Btu/scf)		Volume Fraction		Heat Component (Btu/scf)
Pentane	3709	x	0.128	=	475
Benzene	3601	x	0.216	=	778
Toluene	4284	x	0.018	=	77
Butane	3113	x	0.248	=	772
			Total	=	2102

Since the waste gas heating value is well above the approximate combustion threshold (150 Btu/scf), the final temperature will be a consequence of the amount of excess (quench) air used in the burner. The excess air needed to maintain the flame temperature below 2200°F in this case is approximately 110%. Most burners can operate at excess air levels in this range. However, there is an upper limit on the amount of excess air that can be used in many burners. If the gas is too rich (heating value too high), it may be necessary to add quench air (or water) downstream of the burner to limit the temperature rise.

6.4.4   DIRTY GAS STREAMS

Some waste streams contain a significant quantity of particulates. Others produce combustion products that generate particulate due to inorganic species present in the waste. Under these conditions, the thermal oxidizer is usually oriented to fire the wastes in a downward flow pattern so that the particulate does not collect inside the thermal oxidizer residence chamber. This arrangement is illustrated in Figure 6.9. Some type of drop-out pot or collection system is included at the bottom of the unit to collect the larger particles by gravity. Combustion products then exhaust horizontally, possibly to other downstream particulate collection devices such as an

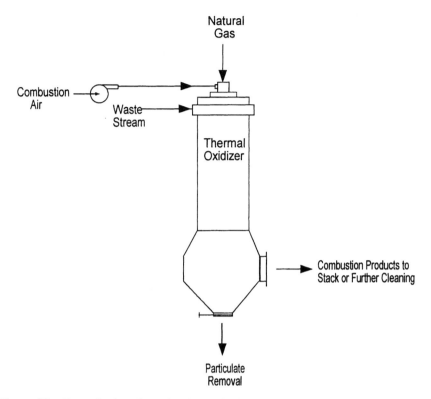

Figure 6.9   Down-fired configuration for particulate.

ESP or fabric filter. Sometimes the particulate is in the liquid or molten phase. Quenching, similar to that shown in Figure 6.1, can be used here, also.

6.5   LIQUID WASTE STREAMS

The focus of this book is the design of thermal oxidation systems for treatment of volatile organic compounds. As such, these VOCs are always part of a gaseous waste stream. If the organic compound is present as a liquid instead of a vapor, it can still be treated using combustion technology. However, it may be classified as a hazardous waste if in the liquid state and be subject to a specific set of treatment regulations. However, not all liquids are classified as hazardous. Those pure compounds that are considered hazardous wastes are listed in the Code of Federal Regulations (40 CFR 261). Nonhazardous liquids are sometimes treated along with VOC-laden waste gas streams. Therefore, liquid waste thermal oxidation will be discussed briefly.

6.5.1   AQUEOUS LIQUID STREAMS

The first category of liquid waste streams is water or wastewater contaminated with organic compounds. These streams have a water content greater than or equal to

85% by weight. Many different industries and processes produce aqueous liquid waste streams. A few examples are

Tank bottoms
Chemical rinses
Process overflows
Process effluents
Storm drainage (runoff)
Landfill leachate
Kraft mill condensate

Since water cannot "burn," it cannot be used as a fuel source. It can be used as a quenching or tempering medium. Water (in the liquid phase) can rapidly quench a flame, causing unstable combustion, high emissions of PICs, nuisance shutdowns, and soot problems. Therefore, the aqueous liquid waste stream should be introduced into the combustion chamber downstream of the burner flame. Liquid injection should be at a point where only hot burner products of combustion can mix with the liquid waste as shown in Figure 6.10.

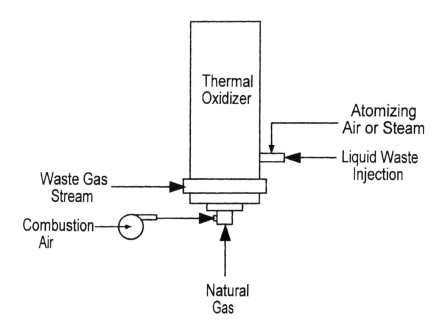

Figure 6.10 Process configuraton for aqueous liquid stream.

The combustion system designer should also assume that the organic compounds in the liquid waste stream do not oxidize until the water is completely vaporized and mixed with oxygen (air). This is a conservative design approach that results in high destruction efficiencies.

The location of the injection point, the method of atomization, and the air/liquid mixing techniques are important factors in optimizing operation, minimizing the chamber size, and maximizing the destruction level. As mentioned above, the aqueous liquid waste stream should be introduced downstream of the burner flame and into a hot gas (POC) stream. A gas temperature of 1800°F or higher is recommended for aqueous waste streams. This high temperature not only ensures high destruction efficiency, it also speeds the rate of liquid droplet evaporation and reduces the overall residence time required for complete destruction.

Unlike VOC-containing gas streams that can be injected directly, liquid wastes must first be atomized. Atomization, typically using air or steam, disperses the liquid into fine droplets which evaporate quickly in the hot environment of the thermal oxidizer. Combustion reactions only occur in the gas phase. Improper atomization can lead to maintenance problems, low destruction efficiencies, and high levels of PICs. Air atomization is preferred, followed by steam atomization for aqueous streams. Mechanical or pressure atomization should only be used when air or steam atomization is not practical. When atomizing with air, the high-pressure air also provides oxygen for oxidation of the organic compounds. However, air atomization should not be used as the primary (or only) method of mixing air with the aqueous waste. The hot burner combustion products generated upstream of the aqueous liquid injection point should also contain excess oxygen.

Typically, final operating temperatures for aqueous liquid waste oxidation systems are higher than most gaseous waste systems. Residence times are also longer in order to allow for droplet evaporation before oxidation can begin. The requirements and features for this process are

- The aqueous liquid stream pressure at the injection point should be 80 to 100 psig
- Air or steam atomization is preferred
- The turndown of the aqueous liquid stream is determined by the injection system design
- The minimum recommended operating temperature (POC outlet temperature) is between 1600 and 1800°F
- The minimum recommended residence time is 1.0 s measured after the aqueous liquid stream injection point
- The aqueous liquid stream injection point is downstream of the burner in an oxygen rich zone

6.5.2 COMBUSTIBLE LIQUID STREAMS

The second category of liquid waste streams is a liquid waste contaminated with a high percentage of combustible compounds. These streams range from 100% combustible compounds to solutions with 30% combustibles and 70% water (by weight). Unlike an aqueous liquid stream, a combustible liquid stream will burn and support stable combustion. Many different industries and processes produce combustible liquid waste streams. A few examples are

Chemical process industries
Petroleum
Petrochemical industry
Pulp and paper
Coating applications
Pharmaceutical industry

A combustible liquid stream should be used as a fuel source. It should be introduced or "fired" through a burner designed for liquids. Environmental regulations may prohibit the introduction of the combustible liquid waste stream until after the thermal oxidizer reaches a minimum operating temperature. For this reason, a dual fuel or multifuel burner with a "conventional" start-up or support fuel is normally supplied as shown in Figure 6.11.

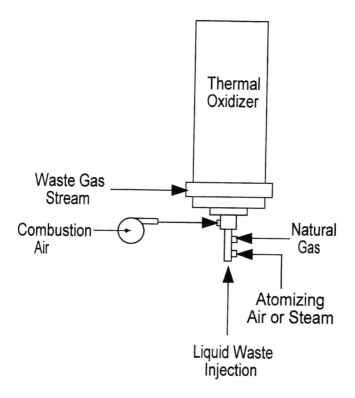

Figure 6.11 Process configuration with multi-fuel burner.

Once again, atomization of the combustible liquid waste is a very important design feature. The same concerns apply as with an aqueous liquid waste stream. Air or steam atomization is preferred. Mechanical or pressure atomization should only be used as a last resort. Operating temperatures and residence times for combustible liquid waste oxidation systems are similar to those for aqueous liquid waste systems. The requirements and features for this process are

- The combustible liquid stream pressure at the burner should be between 20 and 100 psig
- Air or steam atomization is preferred
- The turndown of the combustible liquid stream is similar to an oil-fired burner
- The minimum recommended operating temperature (POC outlet temperature) is between 1600 and 1800°F
- The minimum recommended residence time is 1.0 s

Example 6.7: A thermal oxidizer is designed to treat two VOC-contaminated vapor streams and one aqueous liquid stream with the following flows and compositions:

	Flow in scfm		Flow in lb/hr
	---	---	---
Component	Vapor Stream #1	Vapor Stream #2	Aqueous Liquid
Allyl chloride	11.05	10	0
Dimethylamine	0.32	0	0
Allyl alcohol	4.49	0	49
Carbon dioxide	0.59	0	0
Water vapor	3.86	0	0
Nitrogen	32.42	221.2	0
Oxygen	9.12	58.8	0
Water (liquid)	0	0	155
Total	62	290	204

The thermal oxidizer is operated at 1800°F to ensure good evaporation and destruction of the liquid waste. The heating value of vapor stream #1 is 514 Btu/scf, for vapor stream #2 it is 68 Btu/scf, and is 4008 Btu/lb for the liquid stream. Because of the relatively high heating values of the waste streams themselves, no further auxiliary fuel is needed to reach the specified operating temperature.

The process configuration is shown in Figure 6.12. Waste vapor stream #2 is a tank vent and does not possess sufficient pressure for injection into the thermal oxidizer on its own. A steam eductor is used to provide this motive force. This steam must be included in the mass and energy balance calculations. In this case, because of the relatively high heating value of the wastes themselves, the additional heat load imposed by this steam does not require the addition of auxiliary fuel.

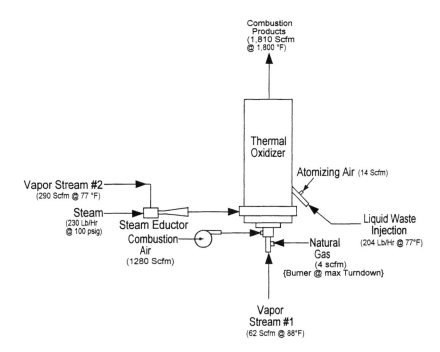

Figure 6.12 Example 6.7 process configuration and flow distribution.

7 Thermal Oxidizer Design

CONTENTS

All thermal oxidizers consist of several minimum basic components. These are a burner, combustion/residence chamber, refractory insulation, and stack. A schematic of a basic thermal oxidizer is shown in Figure 7.1. Of course, most include other components and features, many of which are discussed in detail throughout this book. The burner generates the high temperature flame by combustion of auxiliary fuel such as natural gas or via a rich waste gas if its heating value warrants its injection through the burner. The combustion/residence chamber allows time for the combustion reactions to go to completion. The refractory insulation protects the

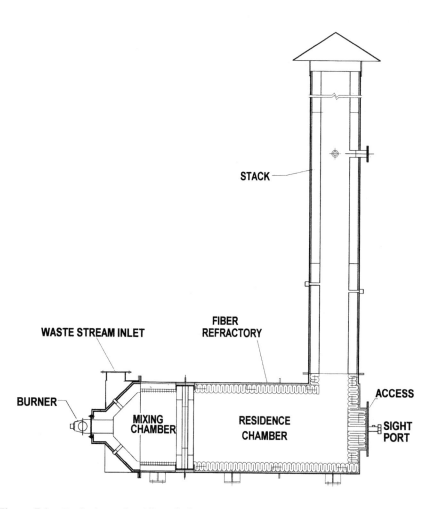

Figure 7.1   Basic thermal oxidizer design.

metal casing from the high temperature gases within. The stack is the conduit for emission of the combustion products to the atmosphere.

7.1 BURNERS

The purpose of the burner is to mix fuel and oxidant (usually air) to produce a flame. This flame provides some or all of the energy required to raise the waste gases to the specified operating temperature of the thermal oxidizer. There are a large variety of commercial burners on the market. Most are designed for specific applications. A thermal oxidizer is one such application.

7.1.1 NOZZLE MIX BURNERS

The primary characteristic of this type of burner is separation of the fuel and oxidant until both exit the orifices of the burner and form a flame upon ignition in the throat of the burner.

Commercial burners are designed so that the leading edge of the flame is retained in a fixed position over a wide range of operation. Usually the air and fuel are mixed in proportions such that the air exceeds the stoichiometric quantity needed to completely combust all of the fuel. The burner is said to operate with "excess air" in this mode of operation. However, there are burners that will operate satisfactorily with as little as 50% of the stoichiometric air required. There are situations where this is advantageous with regards to fuel economy or minimizing NOx (nitrogen oxide) emissions.

A schematic of a simplified, generic burner is shown in Figure 7.2. Auxiliary fuel and air are mixed by a turbulent diffusion process occurring within the flame. This mixing energy is supplied either by pressure drop over the burner (air or fuel), swirl, baffling, or other arrangements. The method by which this mixing energy is generated distinguishes burners and burner manufacturers. Usually the air is injected around a central core of auxiliary fuel. Ignition of the fuel and air occurs downstream of the point where they mix. Burning will only occur where air and fuel are in specific concentration ranges. Only in a thin air/fuel interface does combustion take place. Eddies of fuel and air produce pockets of flame which are carried into unignited zones that create new flame pockets. In this "diffusion" flame, burning is controlled by the rate of mixing between the air and fuel. This involves transport of both mass and momentum between regions of high velocity and regions of low velocity. The flame front remains detached from the gas nozzle since the flame velocity is less than the nozzle velocity.

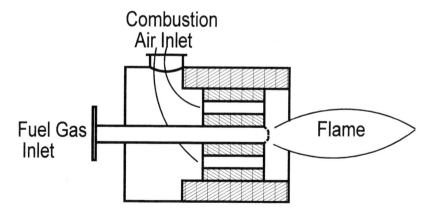

Figure 7.2   Generic burner configuration.

The flame produced has a characteristic length and diameter for a given fuel firing rate and excess air. Combustion air preheat can also affect the flame dimensions, as can the dimensions of the chamber into which the burner is firing. In general, higher excess air rates produce shorter flame lengths. These flame lengths vary with burner design and manufacturer.

If the gas or air velocities through their respective nozzles or orifices are too high, the flame can blow out. Special geometries and baffles are included in the burner design to enhance burner stability. They maintain a localized zone within or adjacent to the flame, which is within the combustible limits of a specific fuel and produces velocities that are low enough to prevent the flame from "blowing-off." One method is to configure the burner tile to slow the air velocity and produce a tangential air swirl to produce recirculation zones which mix hot combustion products with fresh fuel and air. The tile is also heated by exposure to the flame, causing it to glow red and re-radiate heat back to the flame. Other methods use staged air entries, and still others use target plates that produce recirculation zones. Preheated combustion air also improves flame stability and increases the turndown range. The higher the temperature, the greater the effect.

7.1.1.1 Flame Characteristics

The flame interacts with the combustion chamber. The flow patterns within the thermal oxidizer affect the mixing of waste gases with the flame and the formation of recirculation eddies. The hot refractory radiates to the flame and affects its temperature and energy release rate. Thus, the burner flame characteristics depend on the overall design of the thermal oxidizer and its operation.

A wide range of flame patterns can be obtained. The principal variables controlling the flame characteristics are the type of fuel used, the amount of turbulence generated between the fuel and air, and the axial and rotational momentum. Natural gas flames tend to have low luminosity at low excess air. As the excess air increases, the flame becomes more compact, with a bluish-white color. A decrease in the excess air causes the flame to become longer, paler, and less intense (lazier). Unsaturated hydrocarbon (double-bonded compounds, e.g., ethylene) present in the natural gas increase the amount of soot formed and make the flame luminous. When either the combustion air or waste gas are preheated, the flame tends to shorten and become less luminous.

Oil flames are generally luminous. This luminosity increases with heavier oils, which tend to have less hydrogen and generate more soot (unburned carbon particulate). The presence of soot increases the luminosity of a flame. Because of the time necessary to evaporate the fuel droplets and mix the vapor with the air, oil flames tend to be longer than gaseous flames.

7.1.1.2 Turbulence and Jetting

Turbulence produces eddies, which are the means of mixing the fuel with the air or waste gas. While some turbulence is generated in the flame itself, most is the result of velocity differences across the flame. These velocity differences are produced by the pressure drop of air and fuel across burner nozzles and orifices. Since the volume of combustion air is ten times (or greater) more than the fuel gas, most of this energy comes from the combustion air.

The momentum produced by the mass rate of air and fuel injected into the burner produces a jet of gas. This momentum is axial momentum if the air and fuel enter through straight tubes. The axial jet flame produces a long, slowly expanding, conical flame. If turning vanes or air registers are used to produce a swirl, the flame has angular momentum. Burners with swirl are typically used to reduce the flame length and consequently the overall length of the combustion chamber. The shape of the surrounding ceramic burner tile or "quarl" can also affect the pattern of the flame and its recirculation.

The placement of the burner on the combustion chamber should avoid impingement of the burner flame on the refractory lining. Such impingement can cause severe thermal stress within the refractory and premature refractory failure. Therefore, it is important to know the burner flame length over the full range of anticipated operating conditions and design the combustion chamber accordingly.

7.1.1.3 Excess Air

Many burners used in thermal oxidizer applications can operate at excess air levels of 100 to 150%. However, there may be circumstances where it is desirable to use the waste gas as the source of combustion air for the burner. In some cases, the use of the entire volume of waste gas in the burner may exceed the stability limits of the burner. One option is to inject some of the waste gas downstream of the burner directly into the residence chamber. A second option is to use a burner that is designed for very high excess air operation. Such burners do exist. They generally consist of two air entries as shown in Figure 7.3. The rate of air injected through the inner port is equivalent to operation of the burner with 50 to 100% excess air. The remaining air is injected through the outer port. The geometry and mixing patterns are such that a stable flame is first established with the inner air, and the outer air only acts to dilute and cool the combustion products. The design must prevent quenching the flame too rapidly. Otherwise, aldehydes, carbon monoxide, or other pollutant species may be formed.

7.1.1.4 High-Intensity Burners

In some burners, the fuel and air are mixed very rapidly to produce a short flame. This is usually done by high velocities or swirl patterns. The combustion intensity is defined in terms of heat released per cubic feet of burner volume. High-intensity burners have heat releases >250,000 Btu/ft^3. While the short-flame length can reduce the overall length of the thermal oxidizer, the high combustion intensity usually results in higher NOx emissions. These are usually low excess air burners that are unstable at high excess air levels. One such industrial version is shown in Figure 7.4.

7.1.2 Duct Burners

Another type of burner used in thermal oxidation applications is the duct (also called line or distributed) burner. These burners are only designed for gaseous fuels only.

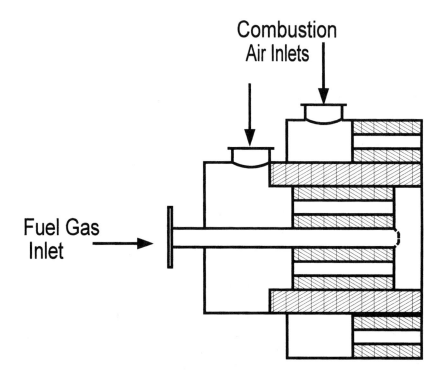

Figure 7.3   High excess air burner with dual combustion air entries.

A schematic of such a burner design is shown in Figure 7.5. Duct burners are used when the fume stream contains a relatively high oxygen concentration and are subject to the limitations described in Chapter 6. Air for combustion is taken from the waste gas stream. Fuel gas enters through holes in manifold pipes placed across the duct. A small flame is formed at the gas manifold hole where the air and gas mix. Unlike the nozzle mix burner that develops one large flame, the duct or line burner produces a multitude of small flames.

Mixing plates are attached to the gas manifold to form a V-shaped pattern. These plates contain holes to allow the air to mix with the gas. Adjustable profile plates are also used to block the area between adjacent burners or between a burner and the wall. The gas flow between the profile plate and the burner mixing plates determines the pressure drop across the burner. It also forces part of the waste gas through the holes in the mixing plate to provide oxygen for the flame.

7.1.3 Low NOx Burners

In response to environmental legislation reducing the emissions of NOx to the atmosphere, an entire class of "low NOx" burners was developed in the late 1980s and early 1990s. The design, operation, and application of these burners are discussed in Chapter 11.

Figure 7.4 High intensity "Vortex" burner. (Courtesy of T-Thermal Company.)

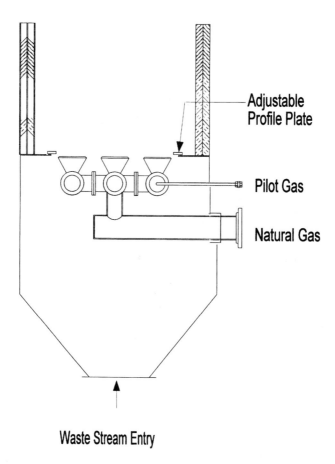

Waste Stream Entry

Figure 7.5   Duct burner schematic.

7.1.4 PREMIX BURNERS

In a premix burner, the fuel and combustion air are mixed before entering the burner proper. This is usually done to lower NOx emissions. The resulting flame temperature is much lower than a nozzle mix burner with the same overall excess air ratio. Care must be taken in the design of these burners to prevent flashback of the combustible fuel/air mixture. This is typically done by using a flame trap (arrestor) or maintaining high velocities through burner orifices.

7.1.5 DUAL FUEL BURNERS

Some thermal oxidizers are designed to operate with either liquid or gaseous fuels. These so-called "dual fuel" burners usually are configured with an oil nozzle down the center of and concentric with the gas nozzle.

    While the use of fuel oil as an auxiliary fuel is not common in the U.S. because of cost and emissions considerations, it is many times used as a back-up fuel. For example, in extreme winter conditions, natural gas is sometimes curtailed to industrial customers to ensure availability to residential customers. In these cases, combustion systems are switched from operation with natural gas to operation with fuel oil using dual fuel burners.

7.1.6 TURNDOWN

The heat content of some waste gases obviates the need for auxiliary fuel. However, the auxiliary fuel burner is rarely turned off. A process upset can temporarily change the waste gas heating value or interrupt flow. Auxiliary fuel firing is then needed and must be available immediately. In circumstances where the waste gas is rich (high heat content), the burner auxiliary fuel-firing rate is usually reduced to its minimum level. This is the level below which it cannot maintain stable combustion. Most nozzle mix burners burning fuel gas (e.g., natural gas) have a turndown of 8:1 to 10:1. For example, if the burner has a maximum firing rate of 10 MM Btu/hr, it can be turned down to 1.0 MM Btu/hr. With high swirl or vortex burners, the turndown tends to be much lower, due to the dependence of mixing patterns on flow rate.

The air turndown range is not usually as wide as the turndown with fuel gas. A typical value is 6:1 for combustion air. Since the fuel turndown and air turndown do not correspond, the excess air rate at turndown will be higher.

Example 7.1: Determine the excess air rate at maximum turndown of a 10 MM Btu/hr burner (max) operating with natural gas at 25% excess air that has a 10:1 fuel turndown and a 6:1 air turndown. Assume the fuel heating value is 1000 Btu/scf.

$$10 \text{ MM Btu/hr} = 10 \times 10^{\wedge}6 \text{ Btu/hr } /1000 \text{ Btu/scf} = 10{,}000 \text{ scf/hr}$$

$$10{,}000 \text{ scf/hr} \times 9.52 \text{ scf air/nat gas (stoichiometric)}$$
$$\times 1.25 \text{ (25\% excess air)} = 119{,}000 \text{ scf/hr burner air}$$

$$\text{At turndown: fuel} = 10{,}000/10 = 1{,}000 \text{ scfh; air} = 119{,}000/6 = 19{,}833 \text{ scfh}$$

$$\text{Excess air at turndown} = [\{(19{,}833/1{,}000)/9.52\} - 1] \times 100 = 108\%$$

The burner turndown when operating with fuel oil is less than gaseous fuels, typically 5:1. Duct (line) burners, which can only operate with gaseous fuels, have turndowns as high as 25:1.

7.2 RESIDENCE CHAMBER

7.2.1 PURPOSE

Combustion is one form of chemical reaction. The products of chemical reactions are governed by two fundamental principles: (1) chemical equilibrium and (2) reaction kinetics. Chemical equilibrium defines the reaction products, assuming the reactants are brought together for an infinite time. Reaction kinetics define the rate at which a reaction occurs. According to chemical equilibrium, a hydrocarbon will react with oxygen to form carbon dioxide and water vapor at 100% conversion. Reaction kinetics do not change the reaction products but define the period of time that the reactants must be in contact for a given degree of reaction to occur. The residence chamber of a thermal oxidizer retains the reactants in contact for a sufficient period of time to satisfy the requirements of chemical kinetics and achieve a high degree of conversion.

The residence chamber is that portion of the thermal oxidizer downstream of the burner and waste gas injection ports. While these various components all comprise the thermal oxidizer, the residence chamber does not begin until all reactants have been injected and thoroughly mixed. Figure 7.6 shows the various components of a basic thermal oxidizer, including the residence chamber. Residence chambers can be cylindrical or flat-sided (rectangular or square). However, flat-sided configurations may suffer from dead spaces where flow is (nearly) stagnant and temperatures vary from the bulk gas temperature. Large flat-sided designs may also require mechanical stiffening to prevent the sides from bulging from pressure fluctuations inside the vessel or from vibrations induced by rotating equipment (e.g., fans) or the burner.

7.2.2 VOLUME SIZING

The need for time, turbulence, temperature, and excess oxygen to achieve high VOC destruction efficiency was discussed in Chapter 4. The residence chamber provides the time. Methods to approximate the residence time required were also described in Chapter 4. The volume of the residence chamber is a function of the residence time required and the volume of combustion products. This volume is calculated by multiplying the volume of combustion products by the residence time required.

$$\text{Chamber volume (V)} = F \times t$$

where F = flow of combustion products (ft³/s)
       t = residence time required (s)

Example 7.2: A waste stream containing 0.63% picoline and 0.34% ammonia in an air stream is thermally oxidized at 1600°F. The combustion products generated are 23,607 acfm, calculated by methods described in Chapter 5. Determine the chamber volume required for a 1.0-s gas residence time.

$$F = 23,607 \text{ acfm} = 393 \text{ ft}^3/\text{sec.}$$

$$t = 1.0 \text{ s}$$

$$\text{Chamber volume} = 393 \times 1.0 = 393 \text{ ft}^3$$

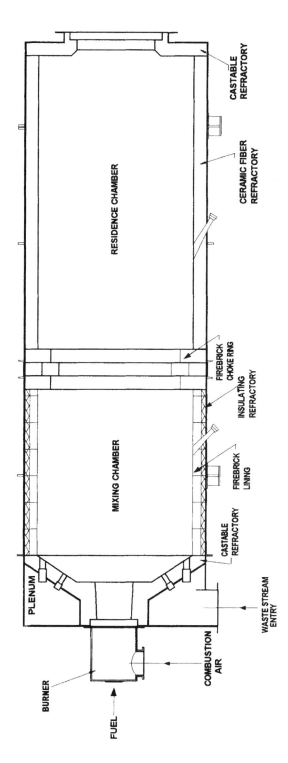

Figure 7.6 Typical thermal oxidizer refractory installation.

7.2.3 DIAMETER

Once the volume has been determined, the diameter must be selected. There are generally two criteria used to select the diameter: (1) length/diameter (L/D) ratio and (2) gas velocity. The L/D ratio is usually in the range of 2 to 8, but can be higher for long residence times. The design gas velocity should be in the range of 25 to 40 ft/s. Remember, if the velocity is in this range at the design (usually maximum) condition, it will be lower at other conditions. Of course, there is a relationship between gas velocity and diameter, and the designer must try to match both to the general design criteria. One procedure for establishing that relationship is as follows:

1. Set the velocity
2. Calculate the diameter
3. Calculate the L/D ratio
4. Repeat steps 1 to 3 until all criteria are satisfied

Useful equations in performing these calculations are as follows:

$$Vel = (flow * L)/V$$

$$D = [(4 \times flow)/(\pi \times Vel)]^{0.5}$$

$$L = (4 \times V)/(\pi \times D^2)$$

where

Vel	= residence chamber velocity (ft/s)
Flow	= flow of combustion products (ft³/s)
V	= residence chamber volume (ft³)
π	= pi = 3.14
L	= residence chamber length (ft)
D	= internal diameter of thermal oxidizer residence chamber (ft)

Example 7.3: Determine the chamber dimensions for Example 7.2.

From Example 7.2, the chamber volume is 393 ft³. Start by setting the velocity to 25 ft/s.

$$D = \{(4 \times flow/(\pi \times Vel)\}^{0.5}$$

$$D = \{(4 \times 393 \text{ ft}^3/\text{s})/(3.14 \times 25)\}^{0.5}$$

$$D = 4.47 \text{ ft (say 4.5 ft)}$$

$$L = (4 \times V)/(\pi \times D^2)$$

$$L = (4 \times 393 \text{ ft}^3)/(\pi \times 4.5^2) = 24.7 \text{ ft}$$

$$L/D = 24.7/4 = 6.2 \text{ (acceptable)}$$

7.2.4 MATERIALS OF CONSTRUCTION

The shell of the residence chamber is typically constructed of carbon steel (ASTM A-36). The thickness is usually 3/16 in. or greater. Sometimes a corrosion allowance is added to a nominal thickness. Stainless steel is not normally required, but units with stainless steel construction are in operation. Stainless steel can be used to provide increased corrosion resistance to acid gases. However, by judicious design of the refractory insulation, the inner metal temperature of the residence chamber can be maintained above the dew point and prevent any corrosion from occurring. Of course, this "hot shell" design will also produce relatively high temperatures on the exterior surface of the thermal oxidizer. Many times a thermal shield (e.g., expanded metal) is offset by 4 to 6 in. from the exterior surface to prevent inadvertent contact by operating personnel. The residence chamber is fabricated of welded or flanged sections to provide a gas-tight construction.

7.3 REFRACTORY INSULATION

The inside of the thermal oxidizer is lined with refractory materials to contain the hot gases and prevent exposure of the metal shell to the high temperature gases. Refractories are generally composed of grains of nonmetallic mineral compositions that are stable, resist softening, and can withstand high temperatures. Typical materials used to make refractories are alumina, bauxite, chromite, dolomite, magnesite, silica, silicon carbide, zirconia, and andalusite. The bonding phase holds the refractory grains together. It is a function of additives to the raw material and subsequent heat treatment. These bonds can be described as a glue phase that totally surrounds the refractory grains and holds them together. The strength of the refractory is dependent on the strength of this phase.

As manufactured, unfired or green refractory masses are often heated or fired in a temperature controlled kiln to develop a ceramic bond between the refractory aggregates. When fired, a high degree of permanent mechanical strength is developed. However, there is an upper temperature limit above which permanent changes in strength, density, porosity, etc. can occur. This upper limit is a critical parameter in the selection of a refractory for application to a thermal oxidizer.

There are four basic types of refractories: (1) brick, (2) castable, (3) plastic, and (4) ceramic fibers. Each is discussed briefly.

7.3.1 BRICK

Refractory brick is a mixture of clays, silica, tabular alumina, or zirconia that have been carefully selected with respect to particle size and composition. This mixture is fired in a kiln to drive off the volatiles and to form a strong ceramic bond between the component particles. Firebrick refers to dense brick, over 100 lb/ft^3, which is normally placed in direct contact with hot combustion gases. Firebrick is also called dense brick. Firebrick generally has relatively poor insulating qualities. It is classified by its composition. The advantage of firebrick is the high temperatures that it can withstand. It can be supplied in a variety of preformed shapes, with $9 \times 4.5 \times 2.5$-in. series and $9 \times 4.5 \times 3$-in. series the most common.

The maximum use temperature of dense firebrick corresponds to the temperature at which a glassy phase within the brick forms, and the brick structure begins to deform. One measure of a refractory's temperature rating is its creep resistance. A refractory's load-bearing strength or creep resistance is determined by the material's subsidence under a compressive load at high temperature.

Table 7.1 shows properties of some common firebrick. Even though the nominal temperature at a particular location in a thermal oxidizer may be measured during operation, inaccuracies in this measurement along with the possibility that higher temperature zones exist in unmeasured locations suggest that a safety factor should be applied when selecting a refractory for a given temperature rating. As a rule of thumb, a 200°F safety factor should be allowed between the maximum expected thermal oxidizer operating temperature and the rated maximum temperature for a particular brick. In fact, this safety factor should be applied to all types of refractory.

TABLE 7.1
Properties of Selected Firebrick

Type	Class	Alumina (wt%)	Maximum Use Temp (°F)
Fireclay	Superduty	40–56	3200
Fireclay	High duty	40–44	3175
Fireclay	Medium duty	25–38	3040
Fireclay	Low duty	22–33	2815
Fireclay	Semisilica	18–26	3040
High alumina	45–48%	45–48	3245
High alumina	60%	58–62	3295
High alumina	80%	78–82	3390
High alumina	90%	89–91	3500
High alumina	Mullite	60–78	3360
High alumina	Corundum	98–99	3660

In contrast to dense firebrick, insulating firebrick (IFB) is a lightweight porous brick, normally less than 50 lb/ft³. Its superior insulating value is a result of its high porosity. The pores, or dead-air spaces, provide a high insulating value. IFB reduces heat loss by two mechanisms: (1) its high resistance to heat flow, and (2) its light weight reduces heat storage. The light weight also facilitates installation. Furthermore, the overall weight of the combustion unit is reduced. It is structurally self-supporting at high temperature. IFB has a low abrasion resistance and should not be used in locations with high gas velocities or gases containing abrasive particulate. Dense firebrick will often be provided with IFB as a back-up between the dense firebrick and metal shell. This provides the benefits of the abrasion resistance and high temperature compatibility of dense firebrick with the high insulating value of IFB. Dense brick is almost always used in mixing chambers and other areas that might experience high velocities or temperatures. This is particularly true around the burner where temperatures are likely to be much hotter than measured at the exit of the thermal oxidizer.

The refractory industry has developed a standard term to describe the size of brick. Called the brick equivalent, or BE, it is a measure of volume. One equivalent typically refers to a brick with dimensions of $9 \times 4.5 \times 2.5$ in. Thus, a brick with dimensions of $9 \times 4.5 \times 3$ in. is 1.2 brick equivalents (3/2.5). One cubic foot of brick is equal to 17.07 brick equivalents. Sometimes the quantity of refractory required is calculated in terms of brick equivalents and is then converted to a particular brick size.

Just as homebuilding bricks must be bonded together to provide a stable construction, the same applies to refractory bricks. Anchors are not used to attach brick to a metal surface. Instead, the bricks are "keyed in" and a thin layer of mortar is used to bond the bricks together. There are many types of mortars available, including air-set, heat-set, and phos-bonded. IFB is typically used as a back-up insulating material and does not see temperatures high enough for a heat-setting mortar.

The most common mortaring practice is to use dipped joints. Here, the mortar is mixed to a thin consistency and the brick dipped into this mixture before being installed. One precaution when using the dipping method is to minimize the amount of mortar on the hot face of the brick. Mortar on this face could cause spalling. Troweled joints are also used, but can result in thicker joints. Both dipped and troweled joints require a skilled brickmason. Remember that refractory bricks are not intended to support structural loads.

When selecting brick for a thermal oxidizer application, consider temperature rating, density, strength (as indicated by modulus of rupture), volume stability, thermal shock resistance, and environment. Volume stability is an indication that the brick will not permanently shrink or expand when exposed to the high temperature gases of the thermal oxidizer. Spalling is the loss of fragments from the face of a refractory. It can be a result of thermal shock, inadequate expansion allowance, or penetration of corrosive gases. Refractory bricks have been developed which are spall resistant.

Brick installation must account for the reversible thermal expansion of the brick as it is heated and cooled. One method is to build in expansion joints. Permanent brick expansion and shrinkage are undesirable properties in a thermal oxidizer application.

7.3.1.1 Physical Properties

There are several physical properties of brick refractories that have an impact on their selection for a given VOC thermal oxidation application. These are apparent porosity, abrasion resistance, cold crushing strength, and modulus of rupture.

Apparent Porosity

Apparent porosity is a measure of the open or interconnected pores in a refractory. The closed pores cannot be readily determined. In any event, it is the open pores that directly affect properties such as resistance to penetration by metals, slags, and fluxes. Generally, refractories with higher porosities have a greater insulating value and better resistance to thermal shock. However, lower porosity improves strength, load-bearing capacity, and corrosion resistance via less slag penetration.

Abrasion Resistance

Abrasion refers to the impact wear of one solid material against another. In thermal oxidation applications, abrasion is usually a result of particles in the waste stream or the formation of particulate during the oxidation process. Greatest abrasion resistance comes from brick that is mechanically strong and well bonded. The modulus of rupture or cold crushing strength can be a good indicator of abrasion resistance.

Cold-Crushing Strength

One of the most widely used parameters for evaluating refractories is strength. It is usually measured at room temperature. Although room temperature measurement cannot be used directly to predict high temperature performance, it is a good indicator of bond formation. This characteristic of a refractory material is determined by applying a compressive load to a sample until the refractory fractures or fails. The cold crushing strength is calculated by dividing the total compressive load applied by the sample cross-sectional area.

Modulus of Rupture

Modulus of rupture is another measure of the strength of a refractory material. It indicates the material's bending or tensile strength at either room temperature or at elevated temperature. In testing conducted to determine this parameter, a sample is supported on both ends in a testing machine and a point load is applied to the center of the sample. The modulus of rupture is expressed in units of lb/in^2 and is the load at which the sample fails.

The physical properties of common refractory brick are shown in Table 7.2.

7.3.2 Castable Refractory

Castable refractories are composed of mixtures of both raw and calcined clays chosen because of their particular composition and particle size range. They also contain compounds, principally calcium aluminate, which hydrate when mixed with water. It is similar to concrete except that it forms strong bonds after heating and dehydration.

Castable refractories are classified as either dense (or hard) or insulating (lightweight). These materials are supplied dry, and water is added for installation. Hydration is the process that bonds the lime in the dry castable with water. The amount of water used is critical in matching the final properties to those specified by the manufacturer. Castable refractory is installed by pouring, troweling, pneumatic gunning, or ramming. As a result, castable refractory provides a continuous, monolithic mass.

Castables are cost effective because they can be installed faster and easier than brick. Mixing, conveying, and placement of the unfired mixture are often mechanized. Brick installation in small spaces can also be time consuming because the brick must be cut to fit.

TABLE 7.2
Physical Properties of Some Common Refractory Brick

Type of Brick	Density (lb/ft²)	Apparent Porosity (%)	Cold-Crushing Strength (lb/in.²⁾	Modulus of Rupture (lb/in²)
		Fireclay		
Superduty	144–148	11.0–14.0	1800–3000	700–1000
High-duty	132–136	15.0–19.0	4000–6000	1500–2200
Low-duty	130–136	10–25	2000–6000	1800–2500
		High Alumina		
60%	156–160	12–16	7000–10000	2300–3300
70%	157–161	15–19	6000–9000	1700–2400
85%	176–181	18–22	8000–13000	1600–2400
90%	181–185	14–18	9000–14000	2500–3000
Silica (superduty)	111–115	20–24	4000–6000	600–1000
		Basic		
Magnesite, fired	177–181	15.5–19.0	5000–8000	2600–3400
Magnesite–chrome, fired	175–179	17.0–22.0	4000–7000	600–800
Chrome, fired	190–200	15.0–19.0	5000–8000	2500–3400
Chrome–magnesite, fired	189–194	19.0–22.0	3500–4500	1900–2300

Source: Courtesy of Harbison-Walker Refractories Company, Pittsburgh, PA.

It is common to add steel fiber to castables to improve toughness and cohesion. For temperatures of 1800°F or lower, these fibers can be #304 or 310 stainless steel. Some castables are made with a silica fume additive. These forms feature easy placement and excellent strength at water contents typically below approximately 6% on a dry weight basis. Strengths are excellent up to 2200°F.

Castables are classified on the basis of density and maximum use temperature. They are available at temperature ratings up to 3200°F. Similar to firebrick, there are insulating castables and dense castables used for more general applications. The temperature limit of a castable is usually stated in degrees. A typical nomenclature might be a lightweight 2700°F castable, meaning that it is lightweight type with a maximum continuous use temperature of 2700°F. The maximum allowable operating temperature of castable refractory is determined by the amount of shrinkage shown at that temperature. This temperature does not consider the environment to which the refractory will be exposed.

The alumina content of any refractory can be usually an indication of its refractoriness. In general, the higher the alumina content, the higher the refractoriness. In castable refractories, the calcium oxide (CaO) can also influence refractoriness.

The iron oxide (Fe_2O_3) content of the refractory can be important for processes with reducing atmospheres. This applies to staged thermal oxidizers designed to minimize NOx formation from chemically bound nitrogen (discussed in Chapter 11). A carbon monoxide atmosphere can react with iron oxide to deposit carbon within the lining. These carbon deposits can eventually cause the lining to crack. Similarly, a hydrogen atmosphere will reduce the Fe_2O_3.

"Drying-out" the castable refractory is the process in which excess water (including the water of hydration) is driven-off before its use. This is done after installation and at a rate and temperature specified by the manufacturer. In cold climates, the refractory must be protected from freezing until the dry-out begins. The time required for dry-out is dependent on the thickness of the lining and generally takes place between 250 and 1000°F. If the drying process occurs too rapidly, it can cause the refractory to spall and crack randomly, which could lead to premature failure. A stronger, ceramic bond is created in some castables when heated to temperatures equal to or greater than 2000°F. Proper dry-out and curing increases the toughness and improves the resistance to abrasion. Predrying the material to remove the water at a controlled rate is necessary to prevent explosive spalling or bursting of the material at high temperatures as the entrapped steam is released. The greatest danger of this occurring comes from dehydrating the hydrated-cement phase during preheat, usually over the range of 400 to 650°F. Explosive spalling can also occur when the castable has frozen in cold weather. It appears to have attained a normal set, but bursts when heated.

Because it will expand and contract during heating and cooling cycles, castable refractory is usually poured or gunned in sections. Sections should be separated by expansion joints to prevent random cracking. Fewer joints are needed compared to a brick lining. Metallic anchors are normally welded to the metal casing to attach the refractory to the inner shell. Refractory is poured or gunned over these anchors.

Castable refractory can also be installed in layers. The resulting composite structure can serve dual purposes: abrasion resistance on the hot face and insulation underneath. In the case of multiple layers, the anchors are sized to retain all layers.

7.3.3 PLASTIC REFRACTORY

This type of refractory material is a mixture of aggregate, clays, and chemical bonding agents that are premixed with water by the supplier. The term "plastic" refers to their elasticity or semisolid form (before curing) which make them amenable to forming using mechanical force. Unlike castables, plastic refractories have water added by the supplier rather than the installer. They are sold in either air-set or heat-set forms. The heat-set variety is divided into two additional types: phosphate-bonded (phosbond) and clay bonded. As their names suggest, the air-set variety hardens on exposure to air, while the heat-set variety hardens on exposure to high temperatures.

Phosphate-bonded refractories develop their strength from reaction of the aggregate in the refractory and the phosphate binder system. They develop the majority of their strength at higher temperatures. This variety is usually preferred in applications where thermal shock or abrasion resistance is required. The clay-bonded and air-set varieties may require exposure to temperatures as high as 1800°F to develop their strength.

Plastic refractories are usually applied by "ramming" into place with pneumatic hammers or special gunning equipment. Again, because the manufacturer adds the water in the proper proportion, they do not suffer from the variability of inconsistent water addition possible with castable refractory. Because of premixed water and chemical bonding, plastic refractories typically have a shorter shelf life than castables. They can dry out during storage or shipment. High ambient temperatures can accelerate this process by driving the binder–aggregate reaction forward and forming a rigid mass. Once hardened, phosphate-bonded refractory cannot be regenerated and must be discarded.

Phosphate-bonded refractories are hygroscopic, meaning they have an affinity to adsorb water from the surrounding air. In high humidity conditions, there have been cases where the refractory slumped. Generally, forms are recommended while heating phosbonded plastics.

Similar to castables, plastic refractories require a controlled heat-up schedule. Typically, preheat rates of 50 to 90°F per hour are acceptable. However, this schedule must come from the material supplier. In some cases, the burner on the thermal oxidizer can be operated at low fire to produce the necessary heat. In others, specialized portable burners or heaters are used to reduce the potential for explosive spalling. Plastics are generally more resistant to thermal shock than castables.

Like castable refractories, plastics require anchors to hold them in place. Anchors may be metal or ceramic pins or various other shapes attached to the metal lining of the thermal oxidizer shell. Every manufacturer and/or installation contractor has a preferred anchoring system and the choice is also very dependent on the application.

After curing, the properties of brick, castables, and plastics of similar composition are almost identical except that the monoliths tend to be slightly less gas permeable than brick. Their ease of application and their uniform thermal and mechanical properties make plastics an ideal selection for making field repairs.

7.3.4 CERAMIC FIBERS

Ceramic fibers are soft, lightweight material made from alumina–silica compositions. They have the advantage of very light weight, very low thermal conductivity, low density, much greater spalling resistance, resistance to thermal shock, ease in application, and low heat storage. Because of these advantages, they are the first choice in the standard VOC thermal oxidation applications where particulate matter, halogen streams, and sulfur are not present. Also, in operations where the thermal oxidizer is started and stopped frequently, the refractory does not deteriorate because of thermal cycling. In addition, there are no heat-up and cool-down schedules to follow. They do not require any special mixing or dry-out procedures like castable and plastic refractories. However, ceramic fibers are much lower in strength and have less resistance to corrosive or abrasive environments. There are also upper limits on the gas velocity of the combustion products. Even though rigidizers can be applied to the surface of the fiber to make it more abrasion resistant, castable refractory is typically used for nozzles and end-walls of thermal oxidizer units that are primarily lined with ceramic fiber.

Refractory fibers are processed into felts, blankets, paper, and block modules. The standard grade has a continuous use temperature of 2300°F. The standard grade products generally have compositions of 47 to 51% alumina and 49 to 53% silica. The alumina/silica ratio within these limits has little influence on thermal resistance. However, the presence of impurities such as sodium (Na_2O) and potassium oxides (K_2O) can have a large effect on the refractory properties. Higher temperature fibers are also available which have a higher alumina content and contain zirconia or other additives. A comparison of different fiber materials and their properties is shown in Table 7.3.

TABLE 7.3
Comparison of Characteristics of Ceramic Fibers

Name	Maximum Service Temp (°F)	Melting Point (°F)	Mean Fiber Diam (micron)	Composition (Wt.%)		
				Al_2O_3	SiO_2	Cr_2O_3
Kaowool	2300	3200	2.8	47.3	52.3	0
Fiberfrax	2300	>3200	2.6–3.0	51.8	47.9	0
Kaowool 1400	2552	3300	2.5	56.3	43.3	0
Fiberfrax H	2606	3506	~2–3	56	44	0
Cerachrome	2600	>3200	3.5	42.5	55.0	2.5
Saffil	2912	3300	3	95	5	0
Fibermax	3000	—	3	72	28	0

Source: Adapted from *Ceramic Fiber Theory and Practice,* Horie, E., Eibun Press, Osaka, Japan, 1986.

The features of ceramic fibers that are the most important in thermal oxidizer applications are fiber diameter and thermal stability. Alumina–silica fibers have diameters ranging from 2 to 3.5 μm (microns). The finer fiber produces a lower thermal conductivity at lower densities and high mean temperatures, but has a marginal effect at higher densities. The maximum thermal limit is established by the manufacturer as the point where the fiber becomes thermally unstable. A phenomenon known as devitrification occurs when this temperature is exceeded. Crystallization begins to occur with consequent shrinkage, loss of strength, and general physical degradation. It is a time/temperature-dependent phenomenon that accelerates as the maximum use temperature is exceeded.

7.3.4.1 Fiber Forms

Ceramic fibers are produced in a wide variety of thicknesses and shapes. Bulk fiber is loose fiber as formed. It is generally used as filler. Original bulk forms can be sprayed with additives to provide flexibility, compressibility, and resilience.

Blanket is a product made of laminated fiber layers in mat form. The layers are set to a specific bulk density. Blanket is manufactured by one of two methods. In one, bulk fiber is scattered by air flow and laminated on a collector net conveyor. No binder is used and the blanket is formed through a thermosetting process. In another method, the bulk fibers are chopped, dispersed in water, and vacuum-formed into mat, either continuously or as a batch operation. Fiber formed in this manner is called **felt**. Ceramic **board** is manufactured using the wet vacuum forming process used to produce felt. The fiber is either bound by inorganic binder or combined with organic binder. As its name suggests, board has little flexibility. Since it is bound with inorganic binders, it maintains its strength even after heating. It does require care against spalling. Board is sometimes used as an expansion joint with brick construction. As brick is installed, a space must be provided for the brick to expand. However, this space must also be insulated to prevent heat from escaping. Using fiber board as an expansion joint provides the necessary spacing while preventing heat from penetrating to the metal shell. Fiber board is rigid enough to provide mechanical strength, yet resilient enough to spring back and refill the joint when the structure cools.

Block, as the name suggests, is a three-dimensional rectangular or square fiber form. It has the same composition and characteristics as blanket. The fiber is oriented thickness wise to have higher strength in that direction when installed. Block linings are installed using either adhesive bonding agents or metal anchors. The metal anchoring method is more common in thermal oxidizer applications. These metal anchors attach the block directly to the inner casing. Many times blanket is folded into an accordion shape (sometimes called stack-bonded) to form a block. There are many anchoring methods used by different manufacturers. One method of ceramic fiber installation is illustrated in Figure 7.7.

Ceramic "**paper**" is a product made from ceramic fiber with an organic binder added. Although it looses its strength after the organic binder is burns out, it can be used as a high temperature gasket or sealing material or as insulating material for small gaps. Ceramic fiber "**rope**" is manufactured by carding (disentangling the fibers with a wire-toothed brush), followed by twisting into yarn. Continuous yarn cannot be manufactured of ceramic fiber alone. Therefore, the ceramic fibers are blended with other fibers such as rayon. Ceramic fiber rope is used as a sealing and packing material.

7.3.4.2   General Construction Techniques with Ceramic Fibers

Layered Construction

Ceramic fiber is generally installed by either layering or modular construction. Layering involves impaling one thickness of insulation on another and attaching to anchors installed on the oxidizer shell. Layers are added until the desired thickness is attained. This method allows the use of various types of ceramic materials in combination.

In the layered construction, a variety of anchor types are available to meet a particular application. Metallic anchors are manufactured in a wide variety of lengths and materials. They can be used for operating temperatures as high as 2200°F. These anchors are welded to the shell and the fibers impaled on them. An anchor washer holds the insulation in place. Ceramic anchors are used when the temperature exceeds 2200°F.

In one method, an alloy metal is welded to the inner shell. The ceramic pin is then threaded onto the metal stud. A ceramic washer then locks on the end of the ceramic pin. The ceramic pin must extend far enough into the refractory lining to prevent exposure of the metal stud to high temperatures. Ceramic anchors are also

❶ Place module in compression against adjacent modules.

❷ Insert patented Pyro-Bloc Stud Gun into end of aluminum tube and nut, partially compressing the spring in the stud gun.

❸ Pull trigger to weld stud, then torque nut on stud, tightening the module and checking the weld. Note: Do not release trigger until both weld sequence and drill sequence are complete.

❹ Remove aluminum tube to ensure it has been rounded out. This assures the weld has been torque tested.

❺ Tamp out lining surface after completion of installation.

Figure 7.7   Ceramic fiber installation. (Courtesy of Thermal Ceramics Inc.)

required in reducing atmospheres that would be encountered in a two-stage thermal oxidizer. The nickel in alloy–metal anchors catalyzes the carbon monoxide to form carbon. This carbon accumulates around stud locations where the temperature is greater than 1000°F and causes fiber deterioration.

Modular Construction

In this construction, prefabricated blocks or modules are manufactured to the correct thickness such that only one layer is required. Generally two methods are used for installing the modules, although there are many manufacturer-specific variations of each. In one method, an embedded metal fitting is attached to the shell by welding or bolting. In the other, separate metal fittings are mounted on the inner casing. Anchor rods, which are part of these metal fittings, are then thrust through the fiber blocks.

7.3.4.3   Surface Coating

Coatings are sometimes applied to ceramic fibers to improve their properties. These coatings, sometimes called rigidizers, can have the following benefits:

- Reduction of shrinkage during high temperature heating
- Increase of surface strength to increase resistance to velocity and abrasion
- Increase in resistance to permeation of combustion gases and resistance to corrosion and scale

7.3.5   CHEMICAL ATTACK

Alumina–silica refractories are relatively inert and resist chemical attack by many materials. However, some components of combustion gases can severely attack refractories, and the selection process must account for them. Like rust on carbon steel, corrosion of refractories involves a chemical reaction between the refractory and constituents of gases surrounding it. Elements or compounds that can attack refractories in VOC thermal oxidation applications are shown in Table 7.4.

Alkali attack from compounds such as sodium, potassium, and calcium oxide (lime) occurs by two mechanisms. Wet alkali attack produces a glazed, glassy surface on the refractory. The refractory is penetrated by the alkali, which reacts to form low-melting eutectic compounds. If the temperature is relatively low, this may not be detrimental. It actually may seal the refractory against further penetration. However, at higher temperatures this phase could melt and drain from the hot face. This exposes fresh surfaces to attack.

Cracking and spalling can result from dry alkali attack. In this case, the alkali penetrates the structure and reacts with the refractory components to form expansive phases containing varied combinations of alkali, alumina, and silica. The reaction product will usually have a different coefficient of thermal expansion from the host refractory. Consequently, cracking and spalling result from mechanical stress, exposing fresh surface to further attack.

Sodium, potassium, and vanadium can react with aluminum oxide and silicon dioxide to form low-melting compounds. Sodium may also attack the alumina portion separately, forming the compound Na_2O-11 Al_2O_3, which tends to promote

TABLE 7.4
Compounds and Elements Potentially
Destructive to Refractories

Name	Chemical Formula
Silicon dioxide (silica)	SiO_2
Ferric oxide	Fe_2O_3
Sodium oxide	Na_2O
Potassium oxide	K_2O
Phosphorus pentoxide	P_2O_5
Lead oxide	PbO
Chloride ion	Cl^-
Fluoride ion	F^-
Vanadium pentoxide	V_2O_5
Calcium oxide (lime)	CaO

bloating and spalling. Vanadium pentoxide and sodium vanadate are found in most low-grade fuel oils.

Calcium, zinc, phosphorus, iron, and cobalt are also often corrosive. They attack refractories via solid state reaction products that are stable at normal operating temperatures but cause spalling during shutdowns. Magnesia (MgO)-based refractories are generally the most tolerant to alkali materials. Magnesia is compatible with alkali and no reaction would be expected. However, magnesia-based refractories do possess some characteristics which could preclude their use. These include a high susceptibility to thermal cycling damage, high thermal expansion rates, and reactivity with halogenated hydrocarbons. Magnesite is also hygroscopic and can result in bloated, expanded refractory. Another choice for alkali resistance would be a high alumina refractory with a dense matrix and low porosity with limited amounts of free alumina and free silica.

Sulfur is both an element of many VOCs (particularly total reduced sulfur [TRS] gases) and fuel oils. In the high temperature environment of the thermal oxidizer, the sulfur reacts to form sulfur dioxide (and small quantities of sulfur trioxide). These sulfur oxides react with refractory compounds to form calcium sulfate ($CaSO_4$), magnesium sulfate ($MgSO_4$), aluminum sulfate ($Al_2(SO_4)_3$), and sodium sulfate (Na_2SO_4). These reaction products weaken the structure of the refractory.

Chlorine gas will also react with the oxides in the refractory to form chlorides of magnesium, iron, or calcium. If a castable refractory is calcium aluminate bonded, calcium chloride can form. If a castable is sodium silicate bonded, sodium chloride can form. Refractories selected should contain minimum amounts of magnesia, iron oxide, and lime. The least reactive oxides are alumina and silica. Therefore, fireclay or high alumina products should be considered.

Fluorine is similar to chlorine. However, fluorine acts aggressively with silicates, while chlorine does not. Fluorine can also react with silica to form silicon tetrafluoride, which is volatile at room temperature. Although fluorine will react with alumina to form aluminum fluoride (AlF_3), this compound also has refractory properties and will not significantly accelerate wear rates. For combustion products containing hydrogen fluoride, refractories should be selected based on minimum silica contents and minimum porosities.

Phosphorous compounds present in a waste stream (e.g., phosphine (PH_3)) will form phosphorus pentoxide in the high temperature oxidizing environment. Phosphorus pentoxide (P_2O_5) reacts with alumina to form highly refractory compounds. In fact, phosphate bonds are commonly used in high alumina refractories. P_2O_5 will also react with silica or magnesia to form low temperature compounds that act as fluxing agents. Refractories chosen for this application should have low silica and magnesia contents.

In two-stage thermal oxidizers designed for NOx control, the first section of the oxidizer operates in a reducing (oxygen-deficient) atmosphere. These conditions tend to reduce any iron oxide in the refractory from the Fe_2O_3 form to FeO. The FeO form is less refractory than the Fe_2O_3 form. This promotes hot-load deformation of the refractory at temperatures above 1800°F. Under reducing conditions, silica (SiO_2) can be converted to the SiO form. High alumina and fireclay refractories with low iron contents, low free silica contents, and low porosities are most suitable for this environment. Air-setting monolithic refractories such as plastics and mortars bonded with sodium silicate and castables with low purity (high iron content) calcium aluminate should be avoided.

7.3.6 TYPICAL REFRACTORY INSTALLATION FOR VOC APPLICATIONS

Figure 7.6 shows a typical refractory installation for a VOC application with no corrosive gases or particulates. The waste stream inlet plenum is lined with a castable refractory to easily fit the geometry of the inlet nozzles and to withstand high temperatures from the burner flame. The mixing chamber consists of a composite with a brick hot face for durability, velocity resistance, and temperature resistance with an insulating backup to minimize heat losses in this zone. This is followed by a brick-lined choke ring for turbulence and mixing. The brick provides mechanical strength against this turbulence and the pressure force resulting from the area reduction. Ceramic fiber is used in the residence chamber to provide good insulating value, light weight, easy installation, and low cost. The flue gas exhaust is again lined with castable refractory to withstand pressure forces from the area reduction and the turbulence created by this reduction.

7.4 THERMAL CONDUCTIVITY

While the refractory selected must withstand the environment into which it is installed, one of the most critical parameters in its selection is its insulating value (resistance to heat transfer). High heat losses through the refractory and ultimately the vessel shell must be offset by a higher auxiliary fuel consumption by the burner. Higher heat losses also produce a higher exterior shell surface temperature that can be hazardous to operating personnel.

The thermal conductivities of various refractory materials are compared in Table 7.5. Thermal conductivity is a function of the mean temperature through the thickness of the refractory. In general, dense brick has the highest thermal conductivity and ceramic fibers have the lowest.

TABLE 7.5
Thermal Conductivities of Selected Refractory Materials
(Btu-in./hr-ft²-°F in an Oxidizing Atmosphere)

	Installed	Alumina	Mean Temperature		
Refractory Type	**Density (lb/ft³)**	**Content (%)**	**500°F**	**1000°F**	**1500°F**
Superduty brick	145	40–45	8.9	9.3	9.7
High alumina brick	183	88	26.5	22.8	20.8
Insulating brick	48	45	1.7	1.9	2.4
Insulating castable	36	66	1.5	1.7	1.9
Gunning castable	121	47	5.5	5.7	5.8
Dense castable	130	45	5.9	6.1	6.3
Plastic	145	65	5.8	6.3	6.7
Ceramic fiber	6	45	0.5	0.95	1.6
	10	45	0.45	0.8	1.2

Note: These refractory materials were selected to be representative of a class of refractories. There can be a considerable variation of properties within a class.

The rate of heat transfer through a flat (slab) refractory wall can be expressed as:

$$Q = [kA \, (T_1 - T_0)]/L$$

where

Q	= heat transfer rate (Btu/hr)
k	= thermal conductivity (Btu-in./hr-ft²-°F)
A	= heat transfer area (ft²)
L	= thickness of material through which heat is transferred (ft)

The area is the surface through which the heat is flowing or for a flat slab equals the width times the length. For a composite (two-layer wall), the equation becomes

$$Q = \frac{(T_1 - T_3)}{(L_1/k_1 A_1 + L_2/k_2 A_2)}$$

where

k_1 = thermal conductivity of material #1 (Btu-in./hr-ft²-°F)
A_1 = heat transfer area of material #1 (ft²)
L_1 = thickness of material #1 (in.)
k_2 = thermal conductivity of material #2 (Btu-in./hr-ft²-°F)
A_2 = heat transfer area of material #2 (ft²)
L_2 = thickness of material #2 (in.)
T_3 = outside surface temperature (°F)
T_1 = inside surface temperature (°F)

These arrangements are shown schematically in Figure 7.8.

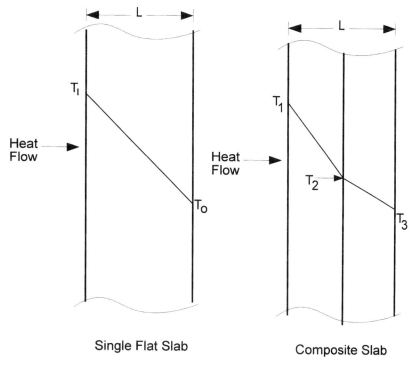

Single Flat Slab   **Composite Slab**

Figure 7.8   Flat wall heat conduction.

For a hollow cylinder, the heat transfer equation becomes:

$$Q = [kA_m(T_i - T_O)]/L$$

where

Q = heat transfer rate (Btu/hr)
k = thermal conductivity (Btu-in./hr-ft²-°F)
A_m = mean heat transfer area (ft²)
L = wall thickness (in.)
T_O = outside surface temperature (°F)
T_I = inside surface temperature (°F)

In this case a mean heat transfer area, A_m, must be used because the inner and outer areas of the cylinder are not equal, due to the wall thickness. The correct equation can be expressed in terms of diameters as follows:

$$Q = \frac{k\pi N \ (D_o - D_I) \times (T_i - T_o)}{\dfrac{(D_o - D_I) \ \ln(D_o/D_I)}{2}}$$

where

N = length of the cylinder (ft)
D_o = cylinder outer diameter (ft)
D_I = cylinder inner diameter (ft)
π = pi (3.14)

For a cylinder with composite walls, the heat transfer equation becomes:

$$Q = \frac{\pi \times N \times (T_I - T_o)}{\dfrac{\ln(D_2/D_1)}{2k_2} + \dfrac{\ln(D_1/D_0)}{2k_1}}$$

where

D_1 = inside diameter
D_2 = outside diameter of inner layer (also inside diameter of outer layer)
D_3 = outside diameter of outer layer
k_1 = thermal conductivity of inner layer (Btu-in./hr-ft²-°F)
k_2 = thermal conductivity of outer layer (Btu-in./hr-ft²-°F)

The refractory layers for single and composite cylinders are shown in Figure 7.9.

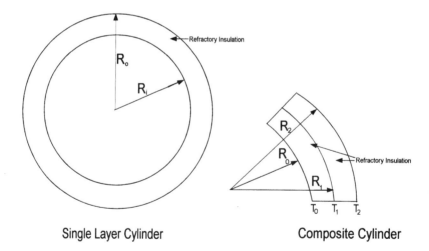

Single Layer Cylinder   Composite Cylinder

Figure 7.9   Cylindrical refractory walls.

7.5   HEAT LOSS

Typically, the heat transfer equations just described are used to determine heat losses from the thermal oxidizer shell. In addition, these calculations can also be used to determine the interface temperature of a composite layer. This is important, since the insulating backup material usually has a much lower temperature rating than the hot face material. Therefore, the interface temperature must be calculated to ensure that this temperature is not exceeded. The thermal conductivity of metals is very high compared with refractory materials. Therefore, it can be assumed that the metal shell temperature is equal to the cold face temperature of the outer refractory layer with very little error. Thus, there is no need to include the metal shell thickness in heat loss calculations.

Example 7.4: A thermal oxidizer is lined with a composite refractory lining of 4.5 in. of dense firebrick and 6 in. of insulating firebrick. The inner diameter of the vessel is 60 in. and the length is 20 ft. The operating temperature is 1800°F and the outer shell temperature is measured as 150°F. Determine (1) the heat loss, and (2) the interface temperature between the two refractory layers. The variation of thermal conductivity with temperature is given below:

	Thermal conductivity (Btu-in./hr-°F-ft²)	
Temperature (°F)	Insulating Firebrick	Dense Firebrick
500	2.1	11
1000	2.4	10
1500	2.8	10
2000	3.4	10

As a first guess, use a thermal conductivity of 2.4 for the insulating firebrick and 10 for the dense firebrick.

$$Q = \frac{\pi N(T_1 - T_o)}{\dfrac{\ln(D_2/D_1)}{2k_2} + \dfrac{\ln(D_1/D_0)}{2k_1}}$$

$$Q = \frac{3.14 \times 20 \times (1800 - 150)}{\dfrac{\ln(81/69)}{2 \times 2.4/12} + \dfrac{\ln(69/60)}{2 \times 10/12}} = 213{,}776 \text{ Btu/hr}$$

The thermal conductivity is divided by 12 to convert to a per-foot basis to be consistent with the units of the other parameters. This heat loss is the same through each refractory layer as it is for the shell. Therefore, the interface temperature can be determined by equating this heat loss to heat transfer for a single layer. Using the dense firebrick dimensions and thermal conductivity:

$$Q = \frac{k\pi N(D_o - D_I) \times (T_1 - T_o)}{(D_o - D_i) \dfrac{\ln(D_o/D_I)}{2}}$$

$$213{,}776 \text{ Btu/hr} = \frac{(10/12)(3.14)(20)\{(69/12) - (60/12)\}(1800 - T_o)}{\left(\dfrac{\{(69/12) - (60/12)\}}{2}\ln(69/60)\right)}$$

Note again that the thermal conductivity is divided by 12 to convert to Btu-ft/hr-ft^2-°F to be consistent with the other parameters that are also per foot.

$$T_O = 1515°F$$

Conversely, the same result is obtained with calculations through the insulating brick layer:

$$213,776 \text{ Btu/hr} = \frac{(2.4/12)(3.14)(20)\{(81/12) - (69/12)\}(T_i - 150)}{\left(\frac{\{(81/12) - (69/12)\}}{2}\ln(81/69)\right)}$$

$$T_I = 1515°F$$

As a first guess in this problem, mean thermal conductivities were used at a temperature of 1000°F for both layers. The actual mean temperature for the dense firebrick is $(1800+1515)/2 = 1658°F$, and for the insulating refractory $(150+1515)/2 = 833°F$. To be accurate, the calculations should be repeated using thermal conductivities at these temperatures. However, in this case, the difference in the heat loss will be small.

There is a fuel cost associated with this 213,776 Btu/hr heat loss. If a single layer of ceramic fiber with a thermal conductivity of 0.9 Btu-in/hr-ft^2-°F could be used in place of the dense and insulating firebrick, the heat loss would be reduced to

$$Q = \frac{(0.9/12)(3.14)(20)\{(81/12) - (60/12)\}(1800 - 150)}{\frac{\{(81/12) - (60/12)\}}{2}\ln(81/60)}$$

$$Q = 51,792 \text{ Btu/hr}$$

The prior discussion and examples addressed heat flow through the refractory lining when either the shell temperature is known or the heat loss is known. In practice, for a new design, neither is known. The shell heat loss or temperature can be determined once the refractory materials and thicknesses have been determined and the thermal oxidizer operating (refractory) temperature has been selected. However, the calculations are more complex and iterative.

The heat loss through the shell must be balanced by heat radiation and convection losses. In equation form:

$$Q_L = Q_R + Q_c$$

where

Q_L = heat loss by conduction through refractory lining
Q_R = heat loss from vessel shell by heat radiation

$$Q_c \quad = \text{heat loss to surrounding atmosphere by convection}$$

The radiation heat loss is determined from the following equation:

$$Q_R = \sigma \varepsilon \, (T_O - T_a)$$

where

σ = Stefan-Boltzman constant (1.713×10^{-9} Btu/ft^2-hr-$^\circ$R^4)
ε = emissivity
T_O = temperature of thermal oxidizer outer shell ($^\circ$F)
T_a = temperature of surrounding air ($^\circ$F)

The convective heat loss to the atmosphere depends on whether this loss is by forced convection or natural convection. Natural convection occurs when no forced air motion occurs. Forced convection occurs when forced air (e.g., wind) moves across the outside surface. For natural convection, the heat transfer depends on the orientation (vertical or horizontal) and geometry (flat or cylindrical walls). For natural convection, one method of calculating the heat transfer is given below.[6]

$$Q_c = 0.53 \times C \times (1/T_{avg})^{0.18} \times (T_O - T_a)$$

where

C = factor accounting for geometry and orientation
T_{avg} = average temperature of surrounding air and oxidizer shell
T_O = temperature of thermal oxidizer outer shell ($^\circ$F)
T_a = temperature of surrounding air ($^\circ$F)

For forced convection, the convective heat loss can be calculated as follows:

$$Q_R = (1 + 0.225 \, \text{vel}) \times (T_O - T_a)$$

where

vel = air velocity (ft/s)
and $T_O - T_a$ as previously defined.

Again, the three heat transfer equations are equated, $Q_L = Q_R + Q_c$ to determine the heat loss or shell temperature. This is done through an iterative procedure. The heat transfer equations are calculated first by assuming the shell temperature is equivalent to the surrounding ambient temperature. A calculated shell temperature is then determined. The heat transfer equations are recalculated with the new shell temperature. These calculations are repeated until there is no change in the calculated shell temperature. Fortunately, these calculations can be performed by the refractory supplier. Some suppliers have developed software to perform these calculations and provide it free of charge to potential customers.

7.6   MIXING

As described previously, VOC destruction efficiency in a thermal oxidizer is a function of residence time, temperature, turbulence, and excess oxygen concentration. Calculation of gas residence time is straightforward, as are measurements of temperature and oxygen concentration. However, turbulence is a difficult parameter to measure. Inadequate turbulence is probably the single most common cause of inadequate VOC destruction in a thermal oxidizer. This causes nonuniform flows and temperatures that allow some of the pollutants to escape without treatment. With odorous compounds such as the TRS gases, ground level concentrations far less than 1 ppm are detectable by the human nose.

Whenever possible, the fume should be injected directly through a burner. This provides the maximum turbulence and mixing and exposes the VOC to the highest temperature. This design concept is applicable when the waste gas stream heating value is greater than 100 Btu/scf or can be used as combustion air for the burner. Other scenarios may demand that the waste stream be injected downstream of the burner.

If the waste stream is a contaminated air stream (i.e., high O_2 content) a duct or line burner may be the best choice, subject to limitations described earlier for this type of burner. It is placed directly in the duct conveying the fume stream and fires toward one end of the residence chamber. To achieve a uniform temperature profile downstream of the burner, the waste stream must flow uniformly into the burner. The typical 1- to 2-in. w.c. waste gas pressure drop across the burner will smooth minor variations. However, when possible, a straight section of 3 to 4 duct diameters upstream of the burner should be included in the design. This is especially important if the waste stream flow rate can fall significantly below the design value.

While mixing the waste stream into the flame zone is required to achieve high VOC destruction efficiency, mixing too rapidly may cause flame quenching. This can produce aldehydes, organic acids, and carbon monoxide that may not oxidize completely in the thermal oxidizer. Both mixing and gas residence time occur within the combustion chamber. One is difficult to distinguish from the other. Usually mixing is a function of pressure drop expended with the injection method used. A balance must be attained between pressure drop energy expended and additional residence time.

With nozzle mix burners, a high velocity jet of flame emanates from the throat of the burner. The waste stream must mix with this jet of flame without quenching the flame. However, simple turbulent mixing is usually not adequate to raise the entire waste stream to the required operating temperature. Burner location, baffles, injection nozzles, and combustion chamber configuration are all used to improve mixing and thermal oxidizer performance.

Mixing is enhanced by high bulk gas velocities through the residence chamber and high L/D ratios which produce eddies. Eddies are produced by adjacent, parallel streams with different velocities. The velocity gradient between the streams generates shear forces which cause a circulating motion called an eddy. Both material and momentum are exchanged between these streams. The process is a form of turbulence. Molecular diffusion becomes important after the eddies have subsided. The chemical reaction between the oxygen and VOC molecule occurs after the final diffusion mixing process has been completed.

To initiate a high rate of mixing, large-scale random eddies should be generated. This is achieved by large velocity differences between the fluids. These high velocities result from pressure energy contained in the waste stream inherently or produced by a mechanical device such as a fan. Jet velocity varies inversely with nozzle area. The distance required to achieve a given degree of mixing varies inversely with this velocity and is a direct function of the energy supplied.

7.6.1   CROSS-STREAM JETS

One approach to mixing in a thermal oxidizer is to provide cross-stream momentum. This is achieved by directing the streams at an angle to each other. Figure 7.10 shows one such arrangement. Here the waste stream is introduced into the burner flame at such an angle as to induce mixing but prevent flame quenching. The velocity and pressure drop across the waste nozzles will influence the degree of mixing.

In a free jet, the turbulent motion at the mixing boundary causes entertainment of surrounding fluid into the flow. The jet spreads and axial velocity falls. One equation for the penetration of a free jet into a quiescent fluid is as follows:[7]

$$y = C_d \, (0.1696 + 2.722 \times (\text{flow})^{0.49} \times (\text{DP})^{0.255} - 0.3975 \times (\text{DP})^{0.255})$$

where

y = radial jet penetration (in.)
C_d = nozzle discharge coefficient (e.g., 0.8)
Flow = flow through nozzle (scfm)
DP = pressure drop across nozzle (in. w.c.)

In a transverse jet, the jet is distorted by the cross-flow. Flow patterns in a transverse jet are illustrated in Figure 7.11. A correlation between transverse jet penetration and axial and radial penetration developed by Patrick[8] is as follows:

$$y/d_O = p^{0.85} (x/d_O)^{\wedge n}$$

where

d_O = jet diameter
y = radial penetration distance
x = axial distance downstream
p = ratio of bulk stream velocity to jet velocity
n = 0.34 for concentration profiles and 0.38 for velocity profiles

This equation applies when $0 < p \leq 0.152$.

For a thermal oxidizer in which a waste stream is injected transverse to the burner flow, the waste stream must be injected to penetrate to at least the centerline

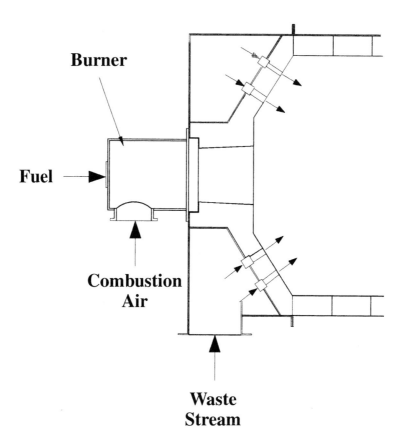

Figure 7.10 Waste stream entry downstream of burner.

of the vessel. The gas residence time should be calculated beginning at the point where this penetration has occurred.

Burners operating with swirl can significantly reduce the distance required for mixing with the waste stream as compared to straight jets. The flame is shortened and becomes larger in diameter. Such a burner can be used to provide hot combustion gases without the danger of flame quenching by the waste stream.

7.6.2   Tangential Entry

As discussed previously for burners, swirl can be used to create intense mixing. Tangential entry in a thermal oxidizer can be used to enhance mixing between the burner flame and the waste stream. Again, pressure drop must be used to produce this swirl pattern.

For best performance, the cold waste stream should be injected axially, while the hot burner flame should enter tangentially. If, instead, the waste stream enters tangentially at high velocity, these cooler (and thus more dense) gases remain near

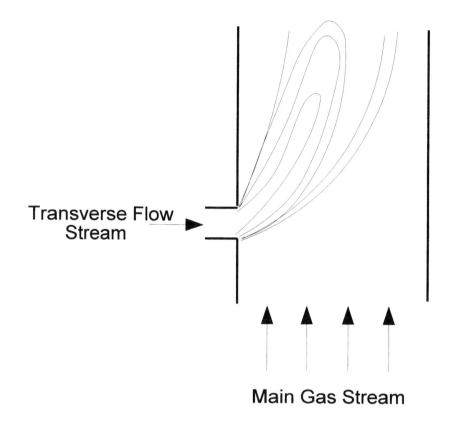

Transverse Flow Stream

Main Gas Stream

Figure 7.11 Effect of main gas stream on transverse jet.

the wall due to the cyclonic motion and mix more slowly with the flame than if injected axially. Tangential firing of the burners is preferred. One such arrangement is shown in Figure 7.12. Mixing can be further improved if the residence chamber expands immediately downstream of the point where the swirl is generated. This produces backflow and cross-mixing. The burner flame should be directed approximately 30° from direct radial injection. Axial entry of the waste stream should be at a relatively low velocity to allow good flame jet penetration. With side-mounted burners, flame impingement on the refractory should be avoided or refractory life may be shortened.

Injecting both the burner combustion products and waste stream with tangential entry can develop a stable recirculating vortex pattern. This pattern could persist over the length of the residence chamber. Therefore, this arrangement should be avoided.

7.6.3 AXIAL JET MIXING

One method of enhancing simple axial jet mixing is to reduce the diameter of the mixing chamber after the flame has been established. The geometry is constructed

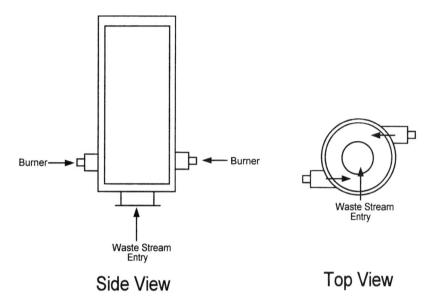

Figure 7.12 Tangentia mixing.

with converging and diverging sections to produce the same affect as an eductor (ejector). The waste stream is then entrained into the hot combustion products of the high velocity flame. This is shown schematically in Figure 7.13.

7.6.4 BAFFLES

Installing baffles is another method of inducing mixing. A baffle is a mechanical obstruction to the normal flow pattern. Baffles serve to block a section of the combustion chamber cross section, resulting in a rapid increase in flow velocity and a corresponding rapid decrease on the downstream side of the baffle. This

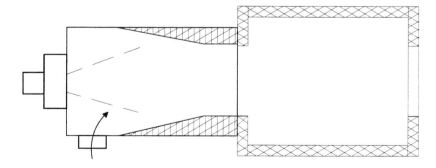

Figure 7.13 Jet-induced mixing.

generates a high level of turbulence and speeds local mixing. Baffles are used in concert with, and not in place of, waste stream injection methods previously described.

A "bridge wall" or "turbulator ring" can be used, depending on the geometry of the thermal oxidizer. With a wall arrangement, symmetry and better mixing are achieved by using two bridge wall segments as shown in Figure 7.14. Baffles should be located to complete mixing as rapidly as possible, allowing the remainder of the residence chamber for completion of the oxidation reactions. To avoid flame impingement, the first baffle should be located 1 to 2 ft beyond the maximum burner flame length. Subsequent baffles should be located approximately one chamber diameter downstream.

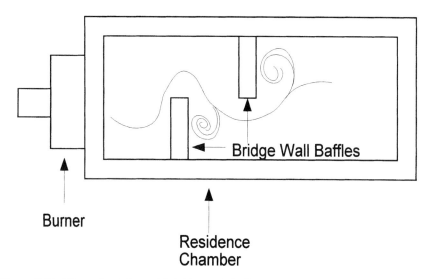

Figure 7.14 Thermal oxidizer with bridge wall baffles.

The baffle walls are usually constructed of refractory brick. The farther the wall extends across the flow area, the better the mixing. However, the farther the wall extends, the greater the pressure drop across the wall. Again, mixing is a trade-off with pressure drop (energy) and the associated costs to develop this pressure drop.

Turbulator or choke rings form a disk at the inner diameter of the residence chamber producing a high velocity throat and backmixing and recirculation patterns downstream of the throat, which enhances mixing. This is illustrated in Figure 7.15. The layer of fluid along the wall of the residence chamber will have a lower velocity than the bulk fluid. These baffles also serve to mix this lower velocity gas with the bulk fluid. In some designs, the same effect is achieved by reducing the shell diameter to form a high velocity throat section, as shown in the photograph of a commercial installation in Figure 7.16.

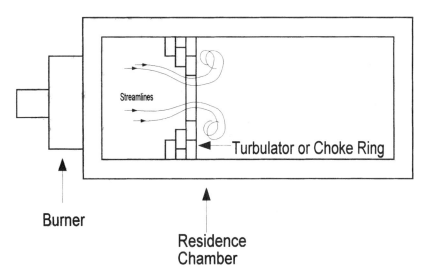

Figure 7.15   Thermal oxidizer with choke ring baffle.

Figure 7.16   Commerical thermal oxidizer with diameter reduction to increase turbulence. (Courtesy of Andersen 2000 Inc.)

7.7   PLENUMS AND NOZZLES

In many designs, the waste stream enters the thermal oxidizer through a plenum with accompanying nozzles. This is illustrated in Figure 7.17. Ideally, the gas velocity through the plenum section should be zero for equal distribution of flow to

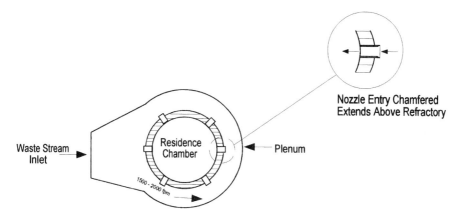

Nozzle Entry Chamfered
Extends Above Refractory

Figure 7.17   Plenum design.

all nozzles. In practice, a velocity of 1500 to 2000 ft/min is acceptable. The nozzles are installed to extend slightly into the plenum. A chamfer on the nozzle entry will reduce pressure drop. The pressure drop across the nozzle can be calculated, using the following formula:[6]

$$\text{Nozzle DP (in. w.c.)} = (v^2/C_d) \times (\rho/0.003)$$

where

$\quad$ V $\quad$ = velocity through nozzle (ft/s)
$\quad$ ρ $\quad$ = density of waste stream at temperature (lb/ft^3)
$\quad$ C_d $\quad$ = discharge coefficient

The discharge coefficient is a function of the design of the nozzle entrance. Discharge coefficients are shown for various configurations in Figure 7.18. A pressure drop of 6 to 8 in. w.c. is typical to achieve good mixing. However, this must be combined with adequate jet penetration, as described earlier.

7.8   TYPICAL ARRANGEMENTS

Due to the wide variety of VOC-containing gases oxidized in a thermal oxidizer, a wide variety of arrangements are in use. Typically, up-fired units are used when no further treatment of the combustion products is required, to conserve plot space, for relatively small units, and when access to thermocouples or other instrumentation at the top of the thermal oxidizer is not critical. Up-fired units are cost-effective because the emission stack can be mounted directly to the outlet of the thermal oxidizer. On the other hand, if a tall stack (based on a flue gas dispersion analysis) is required, costs can escalate because of the steel structure required to support this stack.

$\quad$ Horizontal designs are used if further treatment of the combustion products is required, if access to instrumentation is crucial, a tall stack is required, or if the thermal oxidizer is very large.

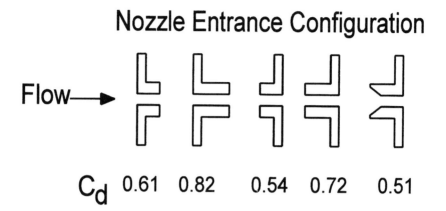

Figure 7.18 Discharge coefficient for various nozzle entrance configurations.

Down-fired designs are typically used when particulate is already in the waste stream or when particulate is produced (generally inorganic) from the combustion reactions. A side gas take-off is located at the bottom of the unit to convey the combustion products to a stack or flue gas treatment device. The disadvantage of this arrangement is that typically a structure must be built around the unit to provide access to the burner at the top.

Many times available plot space will dictate the arrangement used. Both vertical and horizontal U-bend arrangements are in operation, and even combinations of both. As subsets of these configurations, designs are in operation with multiple burners, axial and radial waste stream injection, and rectangular and cylindrical cross sections. Heat recovery options not yet discussed present an entire set of new options, as do certain configurations to minimize NOx reduction which will also be discussed later. While it would seem that only a limited number of designs would be applied to VOC thermal oxidation, in practice there are multiple designs employed by multiple vendors, producing a very wide variety of thermal oxidation systems.

A few of the design concepts discussed are illustrated in Figure 7.19.

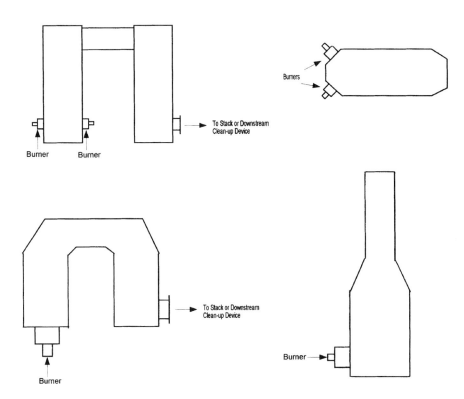

Figure 7.19 Possible thermal oxidizer configurations.

8 Heat Recovery

CONTENTS

Thermal oxidation is a high temperature process. Energy is required to produce this temperature. It can be produced by the VOC oxidation reactions themselves, auxiliary fuel combustion, or a combination of both. The energy contained in the products of combustion can be valuable. It can be used to preheat the waste gas or combustion air or both. It can also be used to produce steam, to heat thermal heat fluids or water, or for drying. One potential drawback of thermal oxidation technology is the operating costs associated with raising the waste stream to the thermal oxidizer operating temperature. These costs can be minimized or even eliminated with heat recovery. In some situations, with steam production, for example, thermal oxidation can be configured to produce a valuable commodity required by a plant.

Heat recovery is not necessary in all VOC thermal oxidation applications. In certain cases, the chemical energy supplied by the VOC thermal oxidation reaction can be sufficient to raise the temperature of the waste stream to the required temperature. Waste streams where the VOC concentration is near or above the lower explosive limit (LEL) fall into that classification.

In most cases, the VOC concentration is too low to sustain combustion. In fact, the VOC concentration of contaminated air streams must be limited to prevent accidental combustion (explosion) at the source. Generally, this limit is set at 25 to 50% of the LEL. The concept of explosibility along with a general discussion of safety is deferred to Chapter 14.

Heat recovery can be classified into two general categories: recuperative and regenerative. In recuperative systems, the heat from the thermal oxidation process exchanges heat via indirect contact with another fluid on a real-time basis. With regenerative systems, heat is stored on an intermediate heat sink material (usually a ceramic solid) for recovery during an alternate cycle. Regenerative systems are discussed in Chapter 10.

As discussed in Chapter 5, both energy and mass must balance in a thermal oxidation process. This means that heat input must equal heat output. The heat input usually consists of one of three forms: (1) chemical energy released from VOCs present in the waste stream; (2) chemical energy from auxiliary fuel added; and (3) sensible heat present when the waste stream, combustion air, or fuel are at an elevated temperature. This is illustrated by the following example.

Example 8.1: A VOC-containing waste stream at 931°F with a flow rate of 10,061 scfm (36,205 lb/hr) is thermally oxidized at 1600°F. The heat content of the waste stream itself is 21.7 Btu/scf (LHV) from the VOCs present. The mean heat capacity of the waste stream between 77°F (reference) and 931°F is 0.36 Btu/lb-°F (due to a relatively high water vapor content). Combustion air is added at a rate of 5507 scfm for VOC oxidation and to produce 5.0% oxygen in the combustion products. Natural gas with a lower heating value of 1110 Btu/scf is added at a rate of 123 scfm (heat release = 8.18 MM Btu/hr). The mass of combustion products produced is 15,818 scfm (61,696 lb/hr), with a mean heat capacity of 0.345 Btu/lb-°F (77° to 1600°F). What is the heat content of the combustion products?

$$\text{Waste stream heat release} = 21.7 \text{ Btu/scf} \times 10,061 \text{ scfm}$$
$$\times 60 \text{ min/hr} = 13.1 \text{ MM Btu/hr}$$

$$\text{Auxiliary fuel heat release} = 8.18 \text{ MM Btu/hr}$$

$$\text{Waste stream sensible heat (at } 931°F) = 36,205 \text{ lb/hr}$$
$$\times 0.36 \text{ Btu/lb-°F} \times (931 - 77) = 11.13 \text{ MM Btu/hr}$$

$$\text{Total} = 13.1 + 8.18 + 11.13 = 32.4 \text{ MM Btu/hr}$$

This heat content can also be calculated directly from the combustion products as follows:

$$\text{POC heat content} = 61,696 \text{ lb/hr} \times 0.345 \text{ Btu/lb-°F}$$
$$\times (1600 - 77°F) = 32.4 \text{ MM Btu/hr}$$

where 61,696 lb/hr is the total mass of combustion products (waste stream + combustion air + natural gas) and 0.345 Btu/Lb-°F is the mean heat capacity of the POC between 77 and 1600°F.

8.1   HEAT EXCHANGERS

One common method of reducing auxiliary fuels costs is to preheat either the waste stream being oxidized, the combustion air, or both. This is done through a heat exchanger(s). The hot combustion products flow through one side (pass) of the heat exchanger, while the waste stream or combustion air flow through alternate passages separated by metal tubes or plates. Heat conveyed to the preheated stream is transferred from the combustion products, resulting in a temperature reduction of the combustion products.

8.1.1   HEAT EXCHANGER TYPES

There are generally two types of heat exchangers used in VOC thermal oxidation heat recovery (recuperative) applications: the plate type and shell and tube. Figure 8.1 shows one embodiment of the plate type. In this illustration, the two gas streams interchange heat in a cross-flow pattern. Combustion products and the heated fluid are separated by individual plates stacked to provide the overall heat transfer required for the intended application. These plate-type exchangers are typically assembled in modules, with each module containing two passes (contact between heated fluid and combustion products in two directions). Increased surface area and heat recovery can be achieved by adding more modules. Plate-type heat exchangers are typically limited to 1550°F flue gas temperatures due to thermal expansion effects.

A schematic of a shell-and-tube heat exchanger is shown in Figure 8.2. Here (usually) the heated fluid flows through individual tubes that are held in place by a tubesheet at each end. Depending on the size of the heat exchanger, intermediate tubesheets may be provided for tube support. The flow pattern between the two gases is also cross-flow, with multiple passes arranged for counter or co-counter flow. A plenum on the backside of the heat exchanger allows the gas to reverse flow directions. This type of heat exchanger can be operated with gas temperatures up to 1800°F with standard materials of construction and even higher temperatures with alloy metals.

In general, plate-type heat exchangers are more cost effective at lower temperatures and when the waste stream and POC are particulate free. The converse is true for shell-and-tube heat exchangers. In some applications, the cost of the heat exchanger may decrease at higher temperatures despite more costly materials because the log mean temperature difference (LMTD) between the gases is greater. Thus, the required surface area decreases along with the overall size of the heat exchanger. The relative capital cost of heat exchangers is shown in Figure 8.3 as a function of effectiveness. This figure only applies to noncorrosive, particulate free gases.

A concern with plate-type heat exchangers is leakage of waste gas between plates and directly to the combustion products. If this occurs, VOC destruction efficiency will suffer, due to the presence of untreated VOC in the stack gas. With shell and tube heat exchangers, tube vibration must also be assessed in the design stage. Vibration of a tube in a tube sheet can cause metal wastage and eventually a hole in the tube where the VOC laden gas will again pass directly to the stack.

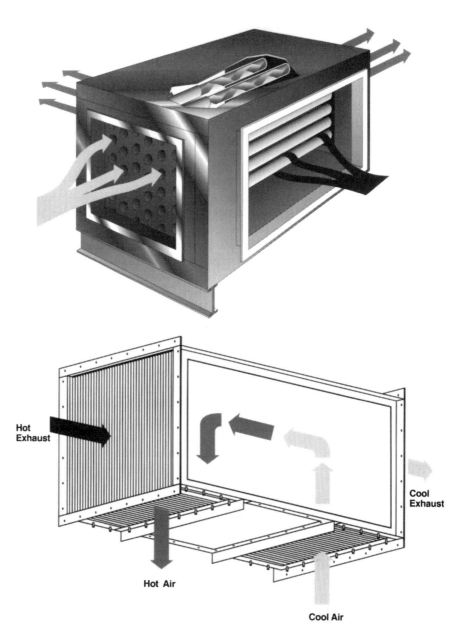

Figure 8.1   (Top) Schematic of shell-and-tube heat exchanger; and (bottom) plate-type heat exchanger. (Courtesy of Exothermics Inc.)

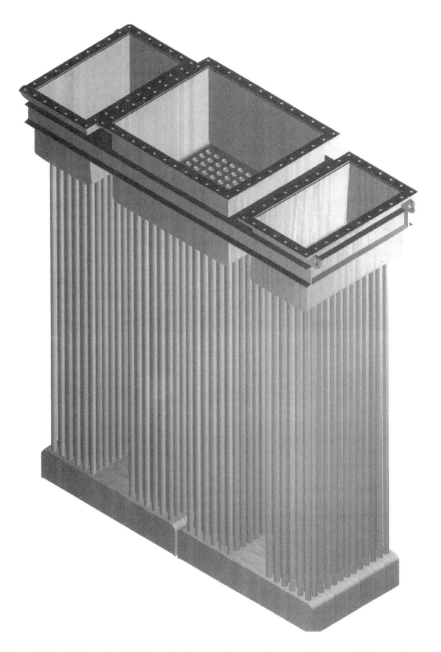

Figure 8.2   Schematic of shell and tube-type heat exchanger. (Courtesy of Alstom Energy Systems SHG, Inc.)

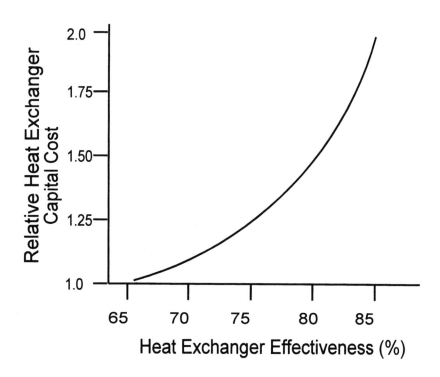

Figure 8.3   Relative heat exchanger capital cost vs. heat recovery efficiency.

Leakage can also result from fatigue failure due to cycling thermal loads. An increase in the stack VOC concentration over time may be indicative of a heat exchanger leak.

8.1.1.1   Heat Transfer

The basic equation for heat transfer between two gases across a heat exchanger is

$$m_1 \times C_{p1} \times (T_{O1} - T_{O2}) = m_2 \times C_{p2} \times (T_{hx1} - T_{hx2})$$

where

m_1 = mass of combustion products (lb/hr)
m_2 = mass of heated fluid (lb/hr)
C_{p1} = heat capacity of combustion products (Btu/lb-°F)
C_{p2} = heat capacity of heated fluid (Btu/lb-°F)
T_{O1} = thermal oxidizer temperature (°F)
T_{O2} = temperature of combustion products exhausting heat exchanger (°F)
T_{hx1} = temperature of fluid entering heat exchanger (°F)
T_{hx2} = temperature of preheated fluid (°F)

Example 8.2: In Example 8.1, the temperature of the waste stream was specified as 931°F. This temperature was achieved by preheating the waste stream by heat exchange with the combustion products. If the temperature of the waste stream before preheating was 190°F, what is the temperature of the combustion products exhausting the heat exchanger?

$$m_1 \times C_{p1} \times (T_{O1} - T_{O2}) = m_2 \times C_{p2} \times (T_{hx1} - T_{hx2})$$

In this case:

m_1 = 61,696 lb/hr
m_2 = 36,205 lb/hr
C_{p1} = 0.345 Btu/lb-°F
C_{p2} = 0.36 Btu/lb-°F
T_{O1} = 1600°F
T_{O2} = unknown
T_{hx1} = 190°F
T_{hx2} = 931°F

$$61,696 \times 0.345 \times (1600 - T_{O2}) = 36,205 \times 0.36 \times (931 - 190)$$

$$T_{O2} = 1146°F$$

The heat transferred is equivalent on each side of the equation. That is, the heat extracted from the combustion products and transferred to the waste stream is exactly equivalent to the increase of the heat content of the waste stream.

$$61,696 \times 0.345 \times (1600 - 1146) = 9.66 \text{ MM Btu/hr}$$
$$= 36,205 \times 0.36 \times (931 - 190)$$

To achieve the desired heat transfer, the heat exchanger must be designed and sized correctly. The temperature gradient between the two fluids will vary along the flow path. The mean temperature difference is calculated from the terminal temperatures. Basic heat exchanger equations relate the total heat exchanged as a function of the total surface area, the overall heat transfer coefficient, and the log mean temperature difference. For a simple countercurrent design, the heat transfer is as follows:

$$Q = U_O A \text{ (LMTD)}$$

where

Q = heat transferred between the two fluids (Btu/hr)
U_O = overall heat transfer coefficient (Btu/hr-ft²-°F)
LMTD = log mean temperature difference between the two fluids (°F)
A = surface area of heat exchanger (ft²)

The log mean temperature difference is defined as

$$\text{LMTD} = \frac{\{(T_{hi} - T_{co}) - (T_{ho} - T_{ci})\}}{\ln\{(T_{hi} - T_{co})/(T_{ho} - T_{ci})\}}$$

where

T_{hi} = initial temperature of hot fluid (°F)
T_{co} = final temperature of cold fluid (°F)
T_{ho} = final temperature of hot fluid (°F)
T_{ci} = initial temperature of cold fluid (°F)

Example 8.3: Combustion products from a thermal oxidizer entering a shell and tube heat exchanger enter the shell side at 1500°F and exit at 800°F. Waste gas flowing through the tubes enters at 100°F and exits at 1000°F. What is the LMTD?

$$T_{hi} = 1500°F$$

$$T_{co} = 1000°F$$

$$T_{ho} = 800°F$$

$$T_{ci} = 100°F$$

$$LMTD = \frac{(1500 - 1000) - (800 - 100)}{\ln\{(1500 - 1000)/(800 - 100)\}}$$

$$LMTD = 594°F$$

Few real heat exchangers are of a simple countercurrent design. Most have several flow reversals or passes to achieve the desired heat transfer. In this case, a correction factor must be applied to the LMTD as follows:

$$\Delta T = F \times LMTD$$

Here, ΔT is now the mean temperature difference to be used in the heat transfer equation. The F factor depends on the flow pattern in the heat exchanger. Values of F for various heat exchanger flow patterns are usually tabulated as a function of the two temperature ratios usually given the symbols P and R. These parameters are defined as follows:

$$P = \{(T_{hi} - T_{ho})/ (T_{hi} - T_{ci}))\}$$

$$R = \{(T_{co} - T_{ci})/ (T_{hi} - T_{ho}))\}$$

Many heat transfer texts and the Tubular Exchangers Manufacturers Association (TEMA) provide figures correlating the F factor with P and R for a variety of heat exchanger arrangements. For example, a shell and tube heat exchanger with 2 shell passes and 4 or more tube passes has a F factor equal to 0.7, when $P = 0.7$ and $r = 1.0$.

The heat transfer coefficient, U_O, is much more complex. It is a function of individual heat transfer coefficients for film resistance for each gas stream, fouling factors, and thermal conductivity through the metal layer separating the gases. For most gas–gas heat exchangers used in thermal oxidation heat recovery systems, the overall heat transfer coefficient is in the range of 5 to 20 Btu/hr-ft²-°F. The design of heat exchangers is best left with vendors of such equipment, who use validated computer programs to perform these calculations.

Example 8.4: A waste stream with a flow rate of 279,000 lb/hr is preheated from 130 to 780°F via exchange with thermal oxidizer combustion products. The mean heat capacity of the waste stream in this temperature range is 0.270 Btu/lb-°F. The operating temperature of the thermal oxidizer is 1550°F and the flow rate of combustion products is 331,800 lb/hr. Heat transfer to the waste stream reduces the temperature of the combustion products to 1047°F. The mean heat capacity of the combustion products between 1550 and 1047°F is 0.293 Btu/lb-°F. The heat transfer coefficient is 8.31 Btu/lb-hr-ft²-°F. What is the heat exchanger surface area?

$$Q = m_1 \times C_{p1} \times (T_{O1} - T_{O2}) = m_2 \times C_{p2} \times (T_{hx1} - T_{hx2})$$

$$Q = 331{,}800 \times 0.293 \times (1550 - 1047) = 279{,}000 \times 0.270 \times (780 - 130)$$

$$Q = 48{,}900{,}000 \text{ Btu/hr}$$

Also,

$$\text{LMTD} = \frac{(1550 - 780) - (1047 - 130)}{\ln\{(1550 - 780)/(1047 - 130)\}}$$

$$\text{LMTD} = 841°F$$

$$Q = U_O A (\text{LMTD})$$

$$48{,}900{,}000 \text{ Btu/hr} = 8.31 \text{ Btu/hr-ft}^2\text{-°F} \times A \times 841$$

$$A = 6997 \text{ ft}^2$$

8.1.1.2 Effectiveness

As a practical matter, all of the high temperature heat contained in one fluid cannot be transferred to another fluid. The proportion of the heat transferred between two fluids as a ratio of the total heat that could be transferred with a heat exchanger with infinite area is termed the effectiveness. For thermal oxidation preheat applications, it can be defined as follows:

$$E = \frac{(T_{co} - T_{ci})}{(T_{hi} - T_{ci})}$$

Once the effectiveness is determined at one set of operating conditions, it can be used to predict temperatures at another set of conditions, assuming that the mass flow rates have not changed. This is done by keeping the effectiveness ratio constant, varying one temperature, and determining the change in a second temperature.

Example 8.5: A waste stream with an initial temperature of 146°F is preheated to 900°F via heat exchange with combustion gases from a thermal oxidizer. The operating temperature of the thermal oxidizer is 1500°F.

1. What is the heat exchanger effectiveness?
2. If the thermal oxidizer operating temperature is changed to 1400°F, what will be the new preheat temperature for the waste stream, assuming all flows remained the same?

1.
$$E = \frac{(900 - 146) \times 100}{(1500 - 146)}$$

$$E = 55.7\%$$

2.
$$0.557 = \frac{(X - 146)}{(1400 - 146)}$$

$$X = 844°F$$

It would seem that the highest heat exchanger effectiveness would be the most desirable in any design. However, the higher the effectiveness the higher the capital cost of the exchanger. Normally, recuperative heat exchangers are used when an effectiveness of 60 to 70% is adequate.

8.1.1.3 Pressure Drop

When using a heat exchanger for recuperative heat recovery, the designer of the thermal oxidation system must account for the additional pressure drop through the system. This includes the pressure drop on both the cold and hot sides. The pressure drop through the cold side accounts for the waste stream flow. The pressure drop through the hot side (combustion products) accounts for the mass of the waste stream, and any combustion air and auxiliary fuel added. Depending on the design and type of heat exchanger used, pressure drops can be as low as 1 in. w.c. to as high as 10 in. w.c. However, higher pressure drops can sometimes be an effective trade-off to less heat exchanger surface area and consequently lower capital costs.

The actual pressure drop must be supplied by the heat exchanger manufacturer and factored into the overall design of a thermal oxidizer. For example, the pressure drop required to supply combustion air to a thermal oxidizer may be 10 in. w.c. without a heat a heat exchanger but could increase to 25 in. w.c. if a combustion air heat exchanger is included in the system. The same is true for the waste stream. Whatever motive force is used to convey the waste stream to the thermal oxidizer must account for a preheat heat exchanger if one is used.

8.1.1.4 Fouling/Coking

One of the terms in the calculation of the overall heat transfer coefficient, (U_O), is the fouling resistance. Fouling can be caused by particulate in the gas stream or condensation of low boiling point organics in the waste stream. Fouling will reduce the overall heat transfer effectiveness. A decrease of waste or combustion air preheat temperature over time could be an indication of fouling.

Fouling can also result from coking. Condensed organic material on the heat transfer surface may encounter temperatures high enough to pyrolyze them. Whereas fouling by particulate may be easy to remove, the coking process causes the material condensed to adhere to the surface almost as a glue. Pyrolysis generally increases as the waste stream preheat temperature increases. Based on experience, preheat of certain types of waste streams is limited to a specific maximum temperature to prevent pyrolysis in the heat exchanger. Waste streams in this category are generally those that contain some oxygen, but at low levels (<5%). Coking and pyrolysis are rarely encountered (except at extreme preheat temperatures) if the waste stream is predominantly a contaminated air stream (high oxygen). In fact, in some applications with contaminated air streams, the waste stream has been preheated to a temperature that exceeds the autoignition temperatures of its VOC constituents without any coking or fouling problems.

If the substance causing the fouling does not adhere too tightly to the surface, it may be cleaned on-line. Sootblowers are one on-line cleaning method. However, they must be generally included in the design stage to allow room for their insertion. Some remain permanently in the gas stream (rotary type) while others are retractable and are only inserted when required. In operation, high pressure steam or compressed air is directed over heat exchanger tubes, dislodging any loosely adhering or brittle substances.

8.1.1.5 Materials of Construction

In most instances, the casing of the heat exchanger must include a refractory lining, just as in the thermal oxidizer. While the tubes or plates have a cooler fluid on one side removing heat, the same cannot be generally said for the shell casing. Because of locally high velocities and turbulence, a durable lining such as castable refractory is used for insulation as opposed to fiber linings.

Metals used in heat exchanger construction are normally Type 304, 309, or 316 stainless steels for metal in contact with hot combustion gases. Type 309 SS usually can be used for metal temperatures up to 1600°F, Type 304 stainless to 1400°F. Type 316 is usually used when dew point corrosion resistance is important. In multipass or module heat exchangers, low alloy steels (such as Corten, P-11, P-22) or carbon steel can be used in cooler stages of the heat exchanger. Corten has a maximum temperature rating of approximately 1000°F, while carbon steel can be used up to approximately 800°F. Again, if the combustion products contain particulate or corrosive gases, recuperative heat recovery may not be an economical design choice.

There are many considerations in the design and application of heat exchangers to VOC thermal oxidation. The design is best left to vendors of such equipment.

In fact, most thermal oxidizer vendors do not design and manufacture heat exchangers themselves but purchase them from heat exchanger vendors when a heat exchanger is included in a system that they are selling. The economics of the heat recovery option must be weighed against fuel savings during the projected life of the system.

8.1.1.6   System Configurations

There are many configurations in which recuperative heat recovery can be used. One is preheat of the waste gas as shown in Figure 8.4. In this case, the waste gas contains a sufficient oxygen concentration to supply the oxygen needed for the burner. Figure 8.5 shows a configuration where the waste gas is oxygen deficient, and both the waste gas and combustion air are preheated. The relative positions of the combustion air and waste gas preheaters could also be reversed. Usually, the stream with the largest flow or highest preheat requirement is heated first. However, there could be reasons to do otherwise. For example, if the waste stream was susceptible to cracking at high temperatures, it would be preferable to preheat the combustion air first. The reduction in the POC temperature through the first heat exchanger would naturally lower the preheat of the waste gas.

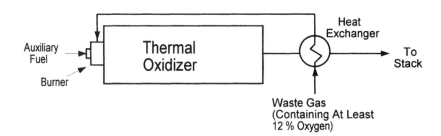

Figure 8.4   Waste gas preheat heat recovery.

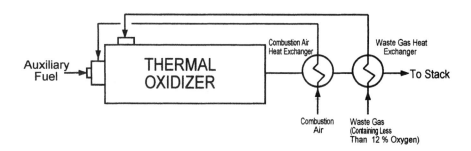

Figure 8.5   Waste gas and combustion air preheat.

Example 8.6: The VOCs in an 1804-scfm (12,409 lb/hr) waste stream (WS) at an initial temperature of 91°F must be thermally oxidized at 1600°F. The waste stream contains 214 lb/hr of carbon monoxide, 1.4 lb/hr of hydrogen, 12.8 lb/hr of methanol, 2.3 lb/hr of methane, and 12,178 lb/hr of carbon dioxide. Determine the auxiliary fuel savings by preheating the waste stream to 1000°F alone and in combination with an 800°F combustion air preheat.

Using methods described in Chapter 5, the combustion products, mean heat capacities, heat exchanger exit temperatures, and auxiliary fuel rates are as follows for the three scenarios:

	Mass Flow Rates (lb/hr)			POC Exit Temperature (°F)		
	Waste Stream	Combustion		Waste Gas Heat Ex.	Combusti on Air HX Heat Ex.	Fuel Rate (MM Btu/hr)
Scenario		Air	Products			
No preheat	12,409	8654	21,383	N/A	N/A	7.2
WS preheat	12,409	4,895	17,439	1038	N/A	3.0
WS and CA preheat	12,409	4,009	16,509	1004	832	2.1

Note that the preheat not only reduces the auxiliary fuel requirements but also reduces the combustion air required, and consequently the mass of combustion products. The combustion air is reduced because (1) less air is needed for the reduced auxiliary fuel rate, and (2) less air is needed to generate 3% oxygen in the reduced volume of combustion products. The mass (and volume) of combustion products is reduced mainly because of the reduced combustion air requirement (the auxiliary fuel reduction also plays a small part). The additional benefit is a smaller residence chamber needed for a specified residence time.

8.1.1.7 Heat Exchanger Bypass

In many VOC thermal oxidizer applications, there can be occasional upward spikes in the VOC loading caused by a process upset. This spike could drive the oxidizer temperature above the setpoint even if the auxiliary fuel valve responds by closing to the maximum turndown rate. One solution to such a scenario is a heat exchanger bypass. This is illustrated in Figure 8.6. Bypass dampers control the flow through the heat exchanger and the bypass. Sometimes only one damper is needed. However, because these dampers are exposed to relatively high temperatures, they are among the most expensive features of this arrangement.

Usually a hot rather than cold bypass is used. The heat exchanger design is dependent to a certain extent on cooling of the plates or tubes of the heat exchanger with the relatively cold waste stream or combustion air. If this cooling medium is removed, metal temperatures could exceed design limits and cause failure. The hot bypass removes the heating medium rather than the cooling medium.

Example 8.7: A recuperative thermal oxidation system is designed as shown in Figure 8.6. The waste stream flow rate is 3787 scfm, with an initial temperature of 248°F and a final temperature of 1100°F after preheating. The normal operating temperature is 1500°F. The maximum temperature rating of the heat exchanger is 1550°F. However, spikes in the VOC loading could drive the temperature above 1600°F if the bypass was not used. How should the system be controlled?

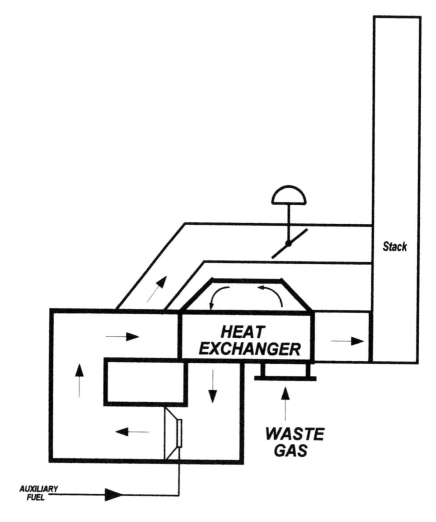

Figure 8.6   Heat exchanger with hot bypass.

Solution

The spikes in the VOC loading are inevitable. The temperature in the oxidizing chamber is controlled to a maximum of 1550°F by modulating the bypass damper to allow some of the flow of combustion products to bypass the heat exchanger. This effectively reduces the preheat temperature of the waste stream and consequently the sensible heat input to the system.

8.2   WASTE HEAT BOILERS (WHB)

Another form of recuperative heat recovery uses waste heat boilers. The boiler and thermal oxidizer are directly connected through a transition section of ductwork.

Hot products of combustion flow through the boiler, generating steam from boiler feedwater. Two types of boilers are used in this application: firetube and watertube. Both units are essentially shell and tube heat exchangers. In a firetube boiler, the hot combustion products flow through a tube bank that is surrounded by water. Conversely, in a watertube boiler, water flows through tubes and hot combustion products flow around the tubes. In contrast to combustion/boiler systems whose sole purpose is to produce steam, waste heat boilers in thermal oxidizer applications are not designed with a radiant section. A radiant boiler section is normally used when a high temperature flame or combustion zone is present. In the thermal oxidizer, such a zone may be present, but it is at the beginning of the thermal oxidizer residence chamber near the burner(s). By the time the combustion products enter the boiler, flames have dissipated.

Firetube boilers are used in smaller applications where the thermal oxidizer heat release is less than 20 MM Btu/hr and when high pressure steam is not required. Watertube boilers are used in larger applications, when high pressure steam is required, or when superheated steam is required. In both boiler types, the typical gas outlet temperature is 350 to 450°F, depending on whether an economizer is used. An economizer, located after the boiler evaporator section, preheats the boiler feedwater before it enters the boiler evaporator section, thus recovering more heat from the combustion gases. Economizers are more common with watertube boilers than firetube boilers. The pressure drop across both types of boilers is typically 3 to 8 in. w.c. Generally, steam pressures are a minimum of 100 psig.

8.2.1 FIRETUBE BOILERS

A photograph of a firetube boiler is shown in Figure 8.7. In this type of boiler design, the high temperature combustion products from the thermal oxidizer flow through the inside of the tube bundle, transferring heat to the pressurized water that surrounds the tubes. The steam produced is discharged to an integral steam drum in the same casing or through riser pipes to a steam drum mounted above the boiler tubes. The single-shell design is less expensive compared to an elevated drum. However, if the gas inlet temperature is high or high purity steam is required, an elevated drum with external downcomers and risers is used. The separate drum enables steam drum internals to be used to achieve the desired steam purity. Single-shell boilers have a small steam space that does not permit installation of steam-purifying equipment.

Firetube boilers can be single pass or multiple pass. In a multiple pass unit, additional passes are added by mounting other drum/tubes arrangements above the first. A plenum at the end of the first pass diverts the flow of combustion products to the next pass. A multiple pass arrangement may be required when space is limited.

Firetube boilers are generally less expensive than watertube boilers for low pressures and capacities. However, their pressure drop is usually higher for the same duty. In general, start-up and response to load changes is slower with a firetube boiler because the tube bundle is submerged in a large body of water. For the same pressure, tube thicknesses are greater in firetube boilers because they are subjected to high external rather than internal pressure. The thickness of the tubesheet also

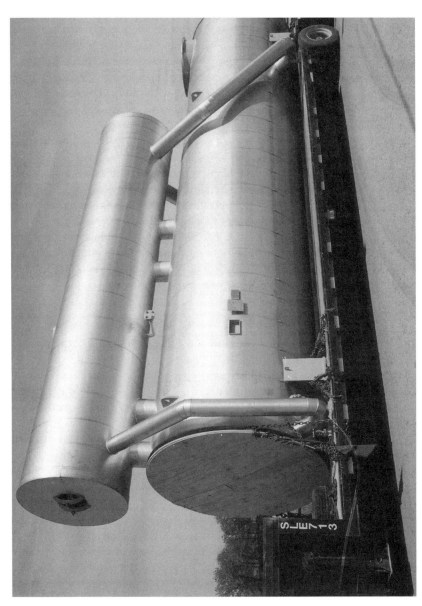

Figure 8.7 Photograph of Firetube boiler. (Courtesy of Rentech Boiler Systems.)

increases with pressure. If the gas stream contains dust or non-adhering materials, a firetube boiler is easier to clean than a watertube boiler. Only the tubes must be cleaned. In a watertube boiler, the boiler casing as well as the tubes must be cleaned. However, if the gas stream contains slagging components, watertube boilers are easier to clean with soot blowers or rapping mechanisms.

8.2.2   WATERTUBE BOILERS

A photograph of a watertube boiler is shown in Figure 8.8. In this design, hot combustion gases flow on the outside of tubes that contain pressurized water inside. A watertube boiler generally consists of tubes rolled or welded into an upper (and sometimes lower) steam drum. A low pressure housing contains the tubes and combustion products. Circulation of the pressurized water can be forced or natural circulation. In most waste heat boilers associated with thermal oxidizers, natural circulation is used. Since the steam is contained inside the tubes, watertube boiler tube thicknesses are less than firetube boilers. Watertube boilers can be designed in a wide variety of shapes and sizes. Some are shop assembled while others must be field-erected. The casing of watertube boilers usually consists of light-gauge sheet metals externally reinforced with structural members. Since thermal oxidizers almost always operate at low pressures, high pressure construction is not required. Like the thermal oxidizer itself, a refractory lining is placed between the boiler tubes and casing of the boiler. This lining is subject to the same selection criteria used for the thermal oxidizer itself, except that gas temperatures are lower.

Extended surfaces such as fins can be added to tubes to enhance heat transfer. However, these are only used with clean gas streams. Water chemistry is important with extended surface designs. High heat fluxes generate higher tube wall temperatures that are more susceptible to fouling from contaminants in the boiler feedwater. Because of the higher heat transfer coefficients associated with gas flow over rather than through the tubes, watertube boilers require less surface area than firetube boilers. Thus, the gas pressure drop can be lower than with a firetube boiler.

8.2.3   ECONOMIZERS

An economizer is basically a gas–liquid heat exchanger located after the boiler evaporator section. Its purpose is to recover additional energy from the combustion products by preheating the boiler feedwater before it enters the steam drum. Water velocity through the economizer tubes is typically 2 to 6 ft/s. Tube sizes usually range from 1 1/2 to 3 1/2 in. If the combustion products contain acid gases, care must be taken with an economizer to prevent metal temperatures from reaching the dewpoint temperature of the acid gas. The calculation of dewpoint temperature is described in Chapter 4. The tube metal temperature is much closer to the feedwater temperature than the gas temperature. One method of avoiding corrosion is raising the feedwater temperature by preheating via a steam to water heat exchanger.

For clean gas streams, fins are sometimes added to the economizer tubes to provide a greater total surface area for heat transfer and enhance energy recovery. The water inside the tubes can absorb heat at a significantly greater rate than the

Figure 8.8   Photograph of Watertube boiler. (Courtesy of Rentech Boiler Systems.)

combustion gases outside the tubes. Finned tubes compensate for the lower heat transfer capacity of the combustion gases. Economizers with finned tubes can reduce the overall size of the economizer by 1/3 to 1/2. Because of the relatively low temperature combustion gases at this point in the system and by maintaining the tube metal temperature above the dew point of any acid gases present, carbon steel tubes can be used. A photograph of an economizer is shown in Figure 8.9.

At high gas turndown conditions, steaming may occur in the economizer. Steaming is the formation of a steam/water mixture that obstructs the flow of the mixture and may result in vibration or water hammer. If operating conditions are anticipated that might produce steaming, the economizer is oriented in an upward direction to aid in the flow and removal of bubbles.

8.2.4 SUPERHEATER

Saturated steam is at a temperature at which the water from which it was generated just begins to boil. Superheated steam is saturated steam whose temperature has been raised further above the boiling point. Naturally, superheated steam has a higher heat content or enthalpy than saturated steam. Some end users, particularly chemical plants, prefer superheated steam to saturated steam.

The location of the superheater in a thermal oxidation waste heat recovery system is usually at the exit of the thermal oxidizer. However, this is not always the case. When the superheater is located upstream of the evaporator, screen tubes are often installed between the thermal oxidizer and the superheater. These are usually large-diameter tubes that act as another evaporator section of the boiler. They serve to reduce the temperature of the combustion gases before they enter the superheater section of the boiler and protect these tubes from over-temperature. In contrast to evaporator or economizer tubes that contain water and thus have a relatively low tube metal temperature, the superheater tubes contain a vapor (steam) and thus their tube metal temperatures are much higher. A variety of superheater, evaporator, economizer configurations are used as shown below:

Evaporator #1 – Superheater – Evaporator #2 – Economizer
Evaporator #1 – Superheater – Evaporator #2
Superheater – Evaporator – Economizer
Evaporator – Superheater – Economizer
Superheater – Evaporator
Evaporator – Superheater
Evaporator – Economizer
Evaporator

The flow path for an evaporator (screen)–superheater–evaporator–economizer configuration is shown in Figure 8.10. Note that the steam drum is the central point for both liquid and vapor flow. This schematic also assumes a natural circulation boiler.

Figure 8.9   Photograph of a Boiler Economizer. (Courtesy of Rentech Boiler Systems.)

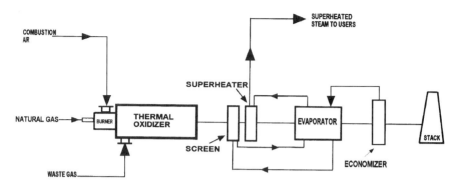

Figure 8.10 Flow path in screen–superheater–evaporator–economizer sections of waste heat boiler.

8.2.4.1 Superheater Temperature Control

In an application with superheat, the superheat temperature could vary widely with the variations in operation or flows to the thermal oxidizer if not controlled. There are several techniques that can be used to control superheat temperature. Two are shown in Figure 8.11. In the first, superheat is controlled by adding water (treated) between superheater sections. In the second technique, the same result is obtained without the use of treated water by condensing part of the steam from the drum and using it to attemporate the superheated steam.

8.2.5 EXTENDED SURFACES

Heat transfer and energy recovery can be enhanced in a waste heat boiler by increasing the surface area. Beyond the tubes themselves, many times "fins" are added to the tubes to increase the effective surface area. These fins can be solid or serrated. Fin densities of 3 to 5 fins per inch are common for the evaporator and economizer. Lower fin densities are used with superheaters. Fins are not used with combustion gases containing particulate.

8.2.6 STEAM DRUM

The steam drum is the central point of liquid and vapor flow in a watertube boiler. Boiler feedwater from the economizer discharges to the steam drum, water flows to and from the evaporator section of the boiler from the drum, and the superheater (if present) discharges to the drum. Steam in the vapor space of the drum then discharges to the steam header. A schematic of a steam drum is shown in Figure 8.12. A combination of cyclones and chevrons in the drum separate the steam from the water as the steam discharges to the header.

    The drum is usually sized for a certain hold-up time of liquid in the drum, similar to the concept of residence time in a thermal oxidizer. This is the duration of time to completely drain the drum if no further feedwater was added. The objective is to ensure that steam can be generated for a short period of time if the boiler feedwater

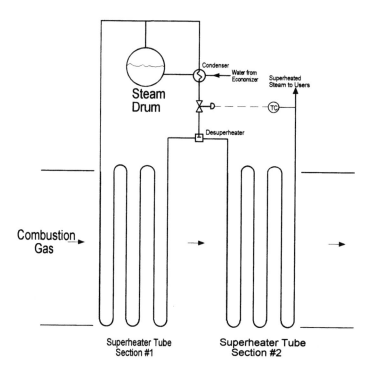

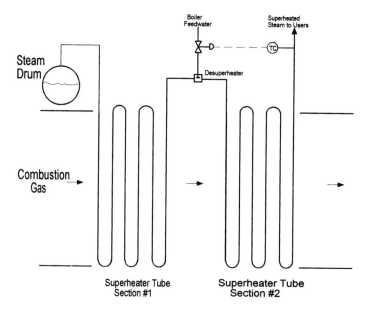

Figure 8.11 Steam superheat temperature control options.

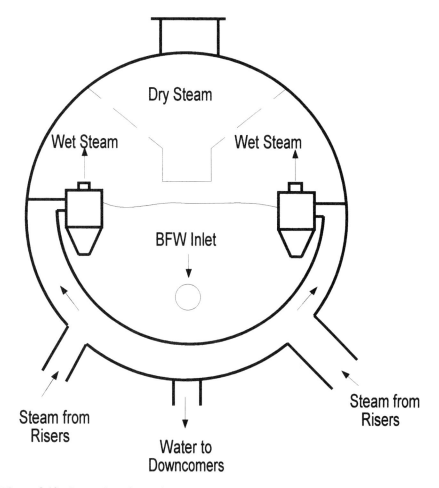

Figure 8.12   Steam drum internals.

pump should stop or the heat source is cut off. This allows a shutdown of the system without causing damage to the equipment. Typical steam drum hold-up times are 3 to 6 min from normal water level to empty.

8.2.6.1   Drum Level Control

Boiler drum level control is a critical parameter in operation of a boiler. A loss of water in the drum could expose the watertubes to heat and stress and eventually damage. The three-element control technique is commonly employed. It is called "three element" because it depends on a response from transmitters of drum level, steam flow, and feedwater flow. The drum level controller manipulates the feedwater flow setpoint in conjunction with feedforward from the steam flow measurement. The feedwater component maintains the feedwater supply in balance with the steam demand. The drum level controller trims the feedwater flow setpoint to compensate

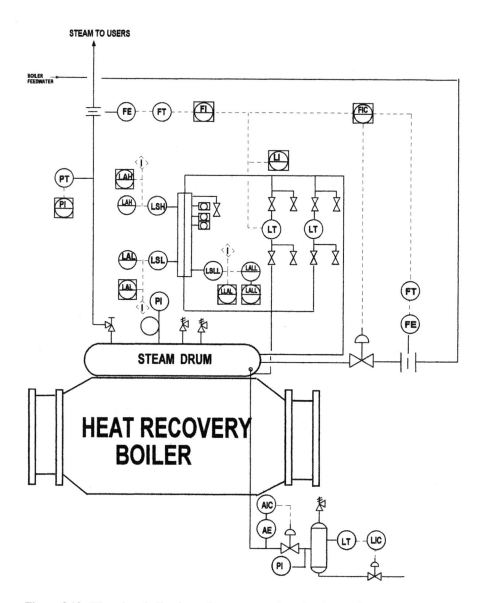

Figure 8.13 Waste heat boiler three-element steam drum level control.

for errors in the flow measurements and other flow disturbances, such as blowdown, that may effect drum level. This control scheme is illustrated in Figure 8.13.

8.2.7 NATURAL CIRCULATION

In a natural circulation boiler, the motive force for steam/water flow through the watertubes (or over the tubes in a firetube design) is the difference in density between cooler water in the downcomers (external to the boiler) and the steam/water mixture in the risers. This flow must be adequate to cool the tubes. In a forced circulation system, a pump is used to ensure the flow of steam and water through the tubes. Natural circulation systems avoid the need for such a pump and thus are preferred in thermal oxidizer waste heat boiler applications. In addition to savings in operating expense, a pump failure would cause a shutdown of the system.

The circulation ratio, CR, is defined as the ratio of the mass of steam/water mixture to the steam generation rate. In waste heat boiler applications, the CR typically ranges from 15:1 for firetube boilers to 20:1 for watertube boilers.

8.2.8 BOILER FEEDWATER

The temperature of the feedwater supplied to a watertube boiler is typically 220 to 250°F. This water is treated before use to control deposition and corrosion in the boiler tubes and associated equipment. Feedwater is really a mixture of returned condensate and treated water. The treated water is commonly called "makeup," because it replaces steam lost by injection into the process or vented. There are two basic types of makeup water: demineralized water and softened water.

Demineralized water is water from which almost all of the contaminants have been removed. The major source of contaminants in the feedwater will be soluble iron and copper from the returned condensate. Softened water is water in which most of the calcium and magnesium ions have been replaced by sodium ions. The feedwater will contain minute quantities of these soluble inorganics, as well as soluble iron and copper from the returned condensate.

Chemical treatment of water for use as boiler feedwater can be divided into three categories: dissolved oxygen control, internal treatment in the boiler, and condensate treatment. Dissolved oxygen must be removed to control corrosion in the feedwater circuit. Most of the oxygen is removed in a deaerator, using steam to drive off oxygen and other dissolved gases. Any oxygen remaining is removed by adding reducing agents called "scavengers" such as sodium sulfite. The oxygen scavenger is fed based on demand that is proportional to the dissolved oxygen concentration.

The objective of internal treatment in the boiler is to control corrosion of the boiler surfaces and deposition of feedwater contaminants. In systems using softened water, this is done by maintaining the pH in the alkaline range by adding chemicals such as caustic soda (sodium hydroxide). To control deposition, various chemicals are used based on precipitation, solubilizing, and dispersing.

Demineralized water in itself is very pure. In boiler systems using demineralized water, the primary feedwater contaminants are primarily iron and copper from the returned condensate. To prevent deposition, a synthetic polymer is added. To prevent corrosion, the pH is controlled.

The purpose of condensate treatment is to minimize corrosion caused by acidic species or dissolved oxygen or both. Three technologies are used individually or in combination: filming amines, neutralizing amines, and passivating agents.

The chemicals selected for treatment are determined by the quality of the feedwater and the boiler's operating pressure. Some chemicals that are used in low pressure boilers are unsuitable for medium and high pressure systems. As the quality of the feedwater varies, some chemicals become more or less effective.

8.2.9 BLOWDOWN

Even with treated boiler feedwater water systems, some small concentration of residual contaminants remain in the water. When the water flashes to steam in the steam drum, these contaminants become concentrated in the remaining water in the drum. To remove these concentrated dissolved solids, a small portion of this water is periodically or continuously removed from the system. This method of removing contaminants is called "blowdown." Blowdown is usually specified as a percentage of the boiler feedwater rate. Typical values are 2 to 5%, but much higher values may be necessary with unusually contaminated water. Blowdown removes water from the system that is at or near the boiling point. The boiler designer must account for this heat loss when estimating the steam production rate.

8.2.10 HOT SIDE CORROSION

Metal corrosion from components in the combustion products can take two forms: high temperature corrosion and low temperature corrosion. High temperature corrosion can be further subdivided into alkali attack and acid attack. Molten alkali salts in a waste stream can be low melting themselves or react with other alkali to form low melting eutectics. These salts or eutectics deposit on tube surfaces and attack the base metal. The coating also insulates the tubes, reducing heat transfer and overall efficiency. Normally, the solution to this problem is to cool the combustion gases below the solidification point of the molten solids before they enter. Typically, this can be done with a water spray, quench air, or recycled flue gas. Sootblowers installed in the boiler can then be used to remove loosely adhering material.

In VOC applications, high temperature corrosion usually occurs due to halogenated compounds in the waste stream. Most common is chlorine in the form of hydrogen chloride in the combustion products. Hydrogen chloride gas is corrosive to metals at metal temperatures above 800°F. This would be most common for a boiler with a superheater. As an approximation, the metal temperature can be assumed to be an arithmetic average of the internal (steam) and external gas temperatures. For example, if the steam is superheated to 800°F and the temperature of the combustion gases is 1600°F, the tube metal temperature of a superheater can be approximated as 1200°F. High nickels steel may be needed to prevent such corrosion.

Low temperature corrosion is usually associated with the economizer, if present, since it represents the lowest gas temperature in the boiler system. Here, dew point corrosion is the main concern and, again, results from halogenated such as chlorine (as HCl). When chlorine is present, the boiler design pressure is usually at least 175 psig to maintain higher steam and consequently tube wall temperatures. Preheat of the boiler feedwater to an economizer may also be necessary to prevent corrosion. Raising the exhaust gas temperature has limited benefit since the liquid temperature has a much greater impact on tube wall temperature of an economizer than does gas temperature.

8.2.11 TEMPERATURE PROFILE THROUGH BOILER TRAIN

The temperature of the combustion products decreases through a waste heat boiler in a more or less uniform rate. At the same time, the boiler feedwater temperature increases (through an economizer, if used), this water flashes to steam in the evaporator, resulting in an abrupt increase in temperature, and the steam temperature increases uniformly in the superheater. The temperature profile through the superheater–evaporator–economizer sections of a waste heat boiler are shown in Figure 8.14. Beginning at the hot end of the system, the temperature of the combustion products is gradually reduced through the economizer, while the temperature of saturated steam is increased through the superheater. In the evaporator section, the temperature of the combustion products continues to decrease, but the temperature of the water/steam mixture on the opposite side of the tubes remains constant as the heat supplied by the combustion gases is used to supply the latent heat required to change the water from the liquid to the vapor state. The difference in temperature between the combustion gases as they exit the evaporator and the water temperature at that same point is called the "pinch" temperature. As the combustion gases proceed through the economizer, their temperature is again reduced, while at the same time raising the temperature of the boiler feedwater. The difference in temperature between the steam saturation temperature and the feedwater as it enters the evaporator section is called the "approach" temperature.

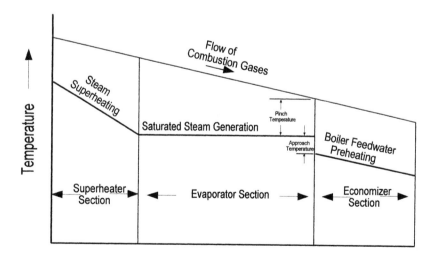

Figure 8.14   Temperature profile through boiler.

The smaller the pinch and approach temperatures, the greater the quantity of steam that will be generated. However, the required surface area of the evaporator will increase as the pinch temperature is reduced. A trade-off must be made between increased steam production and increased evaporator cost. Reducing the pinch temperature also increases the equipment size, resulting in a larger pressure drop through the system.

Care must also be exercised in selecting the approach temperature. If this temperature is too small, "steaming" in the economizer may occur under certain operating conditions. Steaming disrupts water flow through the economizer and is generally considered undesirable.

8.2.12 HEAT RECOVERY ENERGY CALCULATIONS

When specifying a waste heat boiler for a VOC thermal oxidation system, generally the exhaust temperature is specified along with the steam temperature and pressure. Knowing the exhaust temperature simplifies calculations for estimating the steam produced. Basically, the energy removed from the combustion off-gas is equal to the energy used to produce steam.

$$Q_{poc} = Qs$$

$$m_{poc} \times Cp_{poc} \times (T_I - T_O) = m_{bfw} \times Cp_{bfw} \times (t_s - t_o) + (m_{bfw} - m_b) \times LH + m_s \times Cp_s \times (t_f - t_s)$$

where

m	=	mass rate of combustion products (POC), boiler feedwater (bfw), blowdown (b), and steam (s)
Cp	=	heat capacities of combustion products (POC), boiler feedwater (bfw), and steam (s)
T	=	temperature of combustion products in (i) and out (o)
t	=	temperature of boiler feedwater (o), saturated steam (s), and superheated steam (f)
LH	=	latent heat of vaporization of water to steam

In this equation, the heat transferred from the combustion products is used to (1) heat the boiler feedwater to saturation temperature; (2) evaporate the water producing steam; and (3) superheat the steam. The blowdown is subtracted to account for the fact that it is heated to saturation temperature (but is not vaporized) before being removed from the system. The mass rate of steam is the difference between the feedwater rate and the blowdown rate. If only saturated steam is produced, the last term of the equation is neglected.

The right side of the equation is the difference in enthalpy between the boiler feedwater and the superheated steam. If the blowdown is ignored initially (and subtracted later), it is simply a product of the enthalpy difference between the feedwater and steam and the boiler feedwater rate.

$$m_{poc} \times Cp_{poc} \times (T_I - T_O) = \Delta H$$

where ΔH is the enthalpy change between the feedwater and resulting steam.

Example 8.8: Oxidation of a VOC-containing waste stream in a thermal oxidizer produces 100,000 lb/hr of combustion products at 1800°F. How much superheated steam (250 psia, 600°F) will be produced in a waste heat boiler if the economizer exit temperature is 350°F? The boiler feedwater temperature is 230°F. The heat capacity of the combustion products is 0.28 Btu/lb-°F. Ignore boiler blowdown.

$$\text{Heat removed from combustion products } (Q_{poc}) = 100,000 \text{ lb/hr} \times 0.28 \text{ Btu/lb-}°F$$
$$\times (1800 - 350)$$

$$Q_{poc} = 40.6 \text{ MM Btu/hr}$$

From steam tables, the boiler feedwater enthalpy is 198 Btu/lb and the superheated steam enthalpy is 1319 Btu/lb. The enthalpy difference is (1319 − 198) = 1122 Btu/lb. The superheated steam production rate is simply the quotient of the heat removed divided by the enthalpy difference.

$$\text{Superheated steam (lb/hr)} = 40,600,000 \text{ Btu/hr} / (1122 \text{ Btu/lb}) = 36,185 \text{ lb/hr}$$

Steam tables are included in Appendix D for conditions typically encountered in thermal oxidizer waste heat boiler applications.

8.2.13 EQUATIONS FOR STEAM PROPERTIES

Thermal oxidizer mass and energy balance calculations are usually performed with a computer program. Heat recovery calculations can be integrated into these calculations by using the following formulas for steam properties.[9]

$$\text{BFW enthalpy (Btu/lb)} = 1.162 \text{ t} - 1.009 \times 10^{-3} \times t^2 - 46.35$$
$$\times (647.3 - t)^{0.5} - 1.404 \times 10^{-6} \times (1.028)^t + 654.36$$

The temperature in this equation is the boiler feedwater temperature in degrees kelvin. Do not overlook the temperature exponent in the fourth term. A much simpler, although not quite as exact, method is to subtract 32 from the boiler feedwater temperature. The result is the enthalpy in Btu/lb.

The steam saturation temperature for a specified pressure can be calculated as follows, using a method developed by Stoa:[10]

$$\text{Saturation temperature } (°F) = 2.718^{(0.22187 \ln P + 4.76920)}$$

where P = pressure (psia)

For superheated steam, the calculations are more daunting, but still can be performed with a spreadsheet program. The superheated steam enthalpy is[9]

$$H \text{ (Btu/lb)} = 775.596 + 0.63296 \text{ t} + 1.62467 \times 10^{-4} t^2 + 47.3635 \log t$$
$$+ 0.043557 + \{C_7 \times P + 0.5 \times C_4 (C_{11} + C_3 (C_{10} + C_3(C_{10} + C_9 \times C_4)))\}$$

where
$$C_1 = 80870/t$$
$$C_2 = (10^{C_1}) \times (-2641.61/t)$$

$$C_3 = C_2 + 1.89$$
$$C_4 = (P^2)/t^2 \times C_3$$
$$C_5 = 372{,}420/t^2 + 2$$
$$C_6 = C_2 \times C_5$$
$$C_7 = C_6 + 1.89$$
$$C_8 = 0.2187 \times t - (126970/t)$$
$$C_9 = 2 \times C_8 \times C_7 - (C_3/t) \times 126{,}970$$
$$C_{10} = 82.546 - (162460/t)$$
$$C_{11} = 2 \times C_{10} \times C_7 - (C_3/t) \times 162{,}460$$

t = superheated steam temperature (K)

P = superheated steam pressure (atmospheres)

8.2.14 PERFORMANCE AT OFF-DESIGN CONDITIONS

When the design of a waste heat boiler is established at one set of operating conditions, it is often necessary to predict its performance at another. Similar to heat exchanger design, the heat transferred from the combustion gases to the water/steam is equivalent to the product of the temperature difference multiplied by the overall boiler heat transfer coefficient and the boiler surface area. In equation form:

$$Q = m \ Cp \ (T_1 - T_2) = U \ A \ \Delta T$$

where ΔT is the log-mean temperature difference as described previously. Rearranging terms

$$\ln\frac{(T_1 - t_s)}{(T_2 - t_s)} = \frac{U A}{m \ Cp}$$

where

T = gas inlet (1) and outlet (2) temperatures (°F)

t_s = steam temperature (°F)

U = overall boiler heat transfer coefficient(Btu/hr-ft²-°F)

A = boiler surface area for heat transfer (ft²)

m = mass flow of combustion gases (lb/hr)

Cp = heat capacity of combustion gases (Btu/lb-°F)

According to Ganapathy,[11] the overall heat transfer coefficients for firetube and watertube boilers varies as a function of the mass flow of combustion gases as follows:

$$U \propto m^{0.8} \text{ for firetube boilers}$$

$$U \propto m^{0.6} \text{ for watertube boilers}$$

Combining equations

$$\ln\frac{(T_1 - t_s)}{(T_2 - t_s)} = K_1 m^{-0.2} \qquad \text{for firetube boilers}$$

$$\ln\frac{(T_1 - t_s)}{(T_2 - t_s)} = K_2 m^{-0.4} \qquad \text{for watertube boilers}$$

K_1 and K_2 are constants determined from the design condition and then applied at off-design conditions.

Example 8.9: Thermal oxidation of a VOC containing waste gas produces a maximum flow of 100,000 lb/hr of combustion products at 1800°F. These combustion products are directed to a watertube waste heat boiler to produce saturated steam at 250 psig. The temperature of the combustion gases as they exit the boiler is 450°F. Estimate the new boiler exhaust temperature if a change in the waste stream gas flow reduces the flow of combustion products to 75,000 lb/hr at the same inlet temperature.

From the previously described formula or from steam tables, the temperature of 250 psig saturated steam is 406°F. The K_2 factor is first determined at the design condition.

$$\ln\frac{(1800 - 406)}{(450 - 406)} = K_2 (100,000)^{-0.4}$$

$$K_2 = 345.57$$

For the off-design condition,

$$\ln\frac{(1800 - 406)}{(T_2 - 406)} = 345.57(75,000)^{-0.4}$$

$$T_2 = 435°F$$

With this new exit temperature, a new steam rate can be calculated.

This method is generally only applicable to waste heat boilers without an economizer. With an economizer, the exit gas temperature is usually lower than steam temperature. This creates a negative number in the log term that is indeterminate.

8.2.15 WASTE HEAT BOILER CONFIGURATIONS

There are a surprising number of different process configurations applicable to heat recovery using a waste heat boiler for a thermal oxidation application. Since a boiler may consist of several integral components (screen/superheater/evaporator/economizer), several arrangements are possible with these components alone.

In addition, heat exchangers can be integrated with waste heat boilers to expand the possibilities even further. Staged thermal oxidation systems for NOx control add still more options.

Steam can be a valuable commodity in some plant operations. A thermal oxidizer with a waste heat boiler may satisfy part of that demand. In fact, a thermal oxidizer can be designed to produce more heat (and consequently more steam) than is generated by treating the waste stream alone. For example, a burner may only need to fire 10 MM Btu/hr of auxiliary fuel to generate the temperature necessary to treat a waste stream. However, the same system could be easily designed to treat the waste stream while overfiring the burner to produce even more heat and more steam. Both the capacity of the burner and combustion air system must be matched to the ultimate steam demand.

The most straightforward design concept is a waste heat boiler at the gas discharge of a thermal oxidizer, as illustrated in Figure 8.15. This boiler may or may not have an economizer. Again, most firetube boilers do not have economizers. A variation of this concept uses the steam produced to preheat the waste stream and/or the combustion air in steam/gas heat exchangers. This is shown in Figure 8.16. The practical upper limit of preheat temperature with steam/gas heat exchangers is approximately 500°F. Both saturated and superheated steam can be used. The steam can be condensed in these heat exchangers, but the sensible heat of the condensate cannot be recovered. A heat exchanger can be included for preheat in various positions in a boiler train. These options are illustrated in Figure 8.17. While some options produce higher steam production rates or lower thermal oxidizer fuel consumption, there is a trade-off between energy savings and capital equipment cost.

Example 8.10: A thermal oxidizer treating a waste stream produces 460,310 lb/hr of combustion products at 1625°F. These combustion products flow through a waste heat boiler arrangement that includes a superheater, evaporator, and economizer. Steam is produced at 675 psig and 700°F (superheated). The economizer gas exit temperature is 350°F and the boiler feedwater temperature is 240°F. The superheated steam produced is used to preheat the waste stream (396,803 lb/hr at 152°F) and the combustion air (62,941 lb/hr at 77°F). Both are preheated to 500°F. Determine the gross quantity of steam produced and the amount of this steam required to preheat the waste stream and combustion air. Assume mean heat capacities of 0.30, 0.28, and 0.25 Btu/lb-°F for the combustion gases, waste stream, and combustion air, respectively. Also assume that the boiler feedwater blowdown is zero.

Either by using the equations described earlier or from the steam tables, the thermodynamic properties of the steam and boiler feedwater are

Boiler feedwater enthalpy = 208 Btu/lb
Temperature of saturated steam = 501°F
Saturated steam enthalpy = 1202 Btu/lb
Latent heat of vaporization = 712 Btu/lb
Superheated steam enthalpy = 1346 Btu/lb

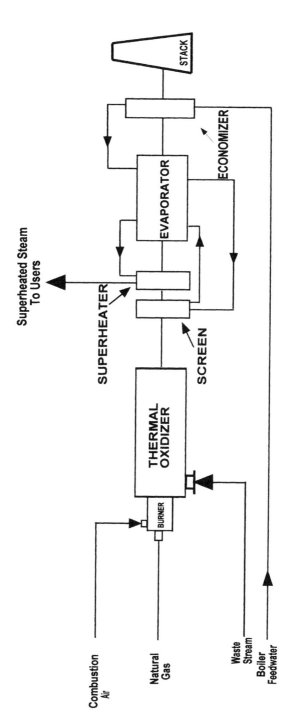

Figure 8.15 Thermal oxidizer with waste heat boiler.

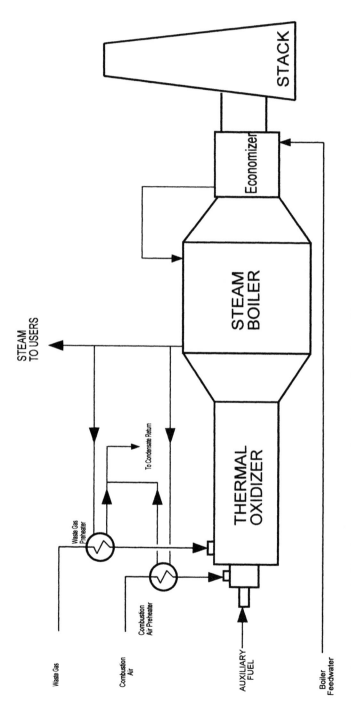

Figure 8.16 Waste heat boiler plus waste stream and combustion air preheat with steam.

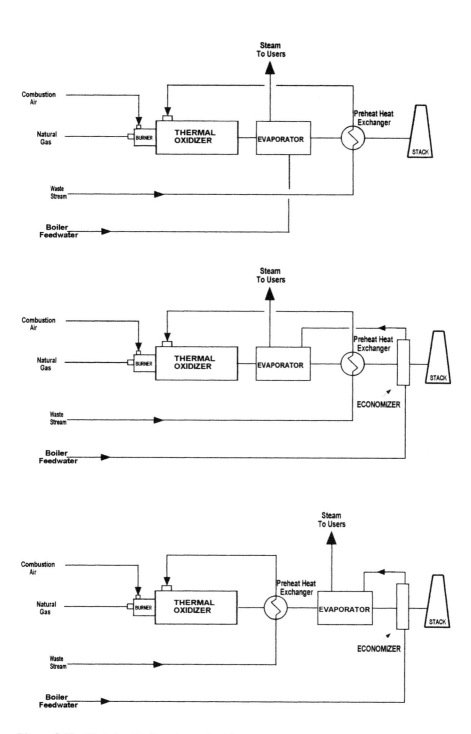

Figure 8.17 Waste heat boiler plus preheat heat recovery concepts.

The heat removed from the combustion products equals the heat used to produce steam.

$$m_{poc} \times Cp_{poc} \times (T_I - T_O) = \Delta H$$

$$460,310 \times 0.30 \times (1625 - 350) = 176,068,575 \text{ Btu/hr}$$

The change in enthalpy of the feedwater to superheated steam is $1346 - 208 = 1138$ Btu/lb. Therefore, the steam generated is

Superheated steam = 176,068,575 Btu/hr/1138 Btu/lb = 154,718 lb/hr

To preheat the waste stream requires the following quantity of heat:

$$Q = m_{wg} \times Cp_{wg} \times (T_I - T_O) = \Delta H$$

$$Q = 396,806 \times 0.28 \times (500 - 152) = 38,664,777 \text{ Btu/hr}$$

The change in steam enthalpy to preheat this gas is $(1346 - 1202) + 712 = 856$ Btu/lb. Therefore, the quantity of steam required to preheat the waste stream is

Superheated steam (lb/hr) = {38,664,777 Btu/hr}/{856 Btu/lb} = 45,169 lb/hr

Using the same procedure, 7,776 lb/hr of superheated steam is required to preheat the combustion air. The net steam available to the plant is $154,718 - 45,169 - 7,776 = 101,773$ lb/hr.

There are several advantages to this particular heat recovery concept. If the plant steam demand increases, the net steam from the thermal oxidation system can be increased by reducing the steam used for preheating the waste stream and combustion air. Also, if the heat exchangers develop leaks, the system could remain in operation until a more convenient time for shutdown and leak repair. In fact, a disadvantage of in-line heat exchangers is that the system must usually be shutdown as soon as a leak occurs or is detected. Again, an increase in the stack VOC concentration can be an indication of a heat exchanger leak with an in-line heat exchanger configuration. Sometimes, only the combustion air heat exchanger is placed in-line. With this configuration, small leaks can be tolerated without any consequence to the process.

8.3   HEAT TRANSFER FLUIDS

In some applications, the plant has no need for steam. Heat transfer fluids present another option. Operation is similar to an in-line heat exchanger, except that the fluid inside is in the liquid state and remains in the liquid state. There are a variety of heat transfer fluids. Some are stable at temperatures as high as 750°F. The most common consist of organic compounds or mixtures of alkylated or benzylated aromatics, hydrogenated and unhydrogenated polyphenyls, and polymethyl siloxanes. Heat transfer fluids are used frequently when the fluid temperature required is 500°F or greater.

Compared to steam, heat transfer fluids require thinner-walled and smaller diameter pipe for a given temperature and heat load. However, leakage is a major concern. This is due not only to the cost of replacement of the fluid but also to the fact that most are flammable. Also, heat transfer fluids tend to be temperature sensitive. The heat exchanger design must account for possible localized hot spots on the tube wall. Typically, the fluid is sampled periodically and checked for thermal decomposition.

Properties of some selected heat transfer fluid are shown in Table 8.1. Figure 8.18 shows a heat recovery configuration with a hot oil heater in combination with a waste stream preheater.

TABLE 8.1
Properties of Selected Heat Transfer Fluids

Fluid Type	Maximum Use Temperature (°F)	Specific Heat (Btu/lb-°F)	Density (lb/ft³)
Diphenyl and diphenyl oxide mixtures	750	0.58	49.3
Di and triarylether mixtures	700	0.54	53.9
Hydrogenated terphenyls	650	0.63	50.3
Aromatic blend	650	0.59	48.7
Alkylated aromatic	600	0.72	35.5
Polyaromatic	600	0.62	49.1
Synthetic hydrocarbon	600	0.70	42.7
Isomeric dibenzyl benzenes	662	0.61	50.85
Isomeric dimethyl diphenyl oxide	626	0.55	49.8
Alkyl diphenyl	707	0.63	48.0
Alkyl benzene	590	0.65	43.7
Dimethyl siloxane	750	0.49	42.0
Polydimethyl siloxane	500	0.54	35.0
50% inhibited ethylene glycol	350	0.93	60.0
50% inhibited propylene glycol	325	0.98	57.1
Paraffinic oils	600	0.70	44.4
Mineral oils	600	0.67	42.4

8.4 WATER HEATING

Similar in concept to the economizer on a boiler, water can be heated to provide hot water for a facility. However, in this case, the combustion gases that enter the water heater are much hotter than an economizer. Care must be taken to be sure that the water does not flash to steam. Sometimes the combustion products are first quenched with air or recycle gas before entering the water heater.

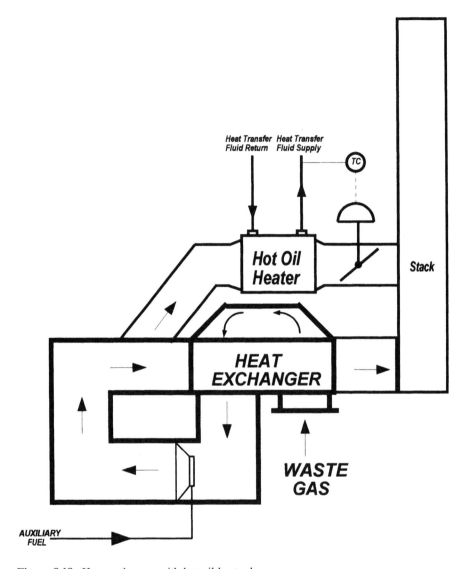

Figure 8.18 Heat exchanger with hot oil heater bypass.

8.5 DRYING

The heat contained in the products of combustion from a thermal oxidizer can also
be used in drying applications. These applications include direct contact drying
(including flash drying) and indirect contact drying. Examples of both are shown in
Figure 8.19.

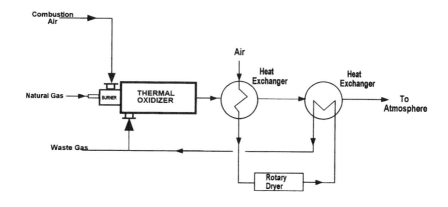

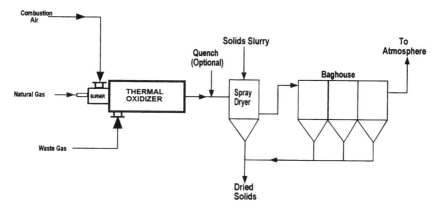

Figure 8.19 Direct and indirect drying concepts.

8.6 REGENERATIVE HEAT RECOVERY

Another form of heat recovery that is periodic in nature is called *regenerative heat recovery.* Here, heat from the combustion products is stored on an intermediate substance, usually ceramic material, for preheat of the waste stream at a later time. Chapter 10 is devoted to a discussion of regenerative thermal oxidation systems.

9 Catalytic Oxidation

CONTENTS

A catalyst is a substance that can change the rate of a chemical reaction but remain unaltered by that reaction. Though a catalyst may speed up a reaction, it never changes the reaction products. Catalysts can be used to oxidize volatile organic compounds (VOCs) at temperatures much lower than required with thermal oxidation. Catalyst materials include platinum, platinum alloys, copper oxide, chromium, cobalt, vanadium, and manganese oxide. These materials are plated in thin layers on inert substrates designed to provide maximum surface area between the catalyst and the VOC-laden gas stream.

9.1 APPLICATIONS

Catalytic converters have been used to control hydrocarbon emissions from automobile tailpipes since 1975. VOC control using catalytic oxidation began in the late 1940s. It was used then primarily for odor control. The Clean Air Act amendments of 1970 along with the energy crisis of 1973 provided the impetus to improve catalytic oxidation technology to be more universally applicable. The chemical industry was the first to utilize catalytic oxidation extensively for emissions control. It has also been used to treat waste streams from automobile paint-baking applications with flows as high as 500,000 scfm. While catalytic oxidation systems have been used to treat very large waste streams, typical applications are less than 5000 scfm. A partial list of applications is shown in Table 9.1.

TABLE 9.1.
Catalytic Oxidizer Applications

Application	VOC Treated
Can coating	Methyl isobutyl ketone, mineral spirits, isophorone, butyl cellosolve
Metal coating	Methyl ethyl ketone, methyl isobutyl ketone, toluene, iso-butanol
Automobile paint baking	Methyl ethyl ketone, toluene, xylene
Glove manufacture	Formaldehyde, phenolics
Phthalic anhydride manufacture	Phthalic anhydride, maleic anhydride
Flexible packagings	Acetates, alcohols, ketones
Pharmaceutical pill coating	Alcohols
Vinyl coating	Ketones, aromatics
Phenol manufacture	Cumene, acetone
Film coating	Ketones, acids
Formaldehyde manufacture	Methyl ether, carbon monoxide, methanol, formaldehyde
Aseptic packaging	Alcohols, acetates
Printing	Ink oils
Electronic components	Methyl ethyl ketone, cellosolve

Source: *Catalytic Control of VOC Emissions — A Guidebook*, Manufacturers of Emission Controls Association (MECA), Washington, D.C., 1992. With permission.

Usually, catalytic oxidizers are not used unless the VOC concentration of the waste gas stream is less than 25% of the lower explosive limit (LEL). The LEL concept is described further in Chapter 14. This represents an overall waste stream heating value of approximately 10 to 20 Btu/scf (LHV).

9.2 THEORY

In solid catalyzed reactions, the presence of catalyst surface in the proximity of reactive gas molecules promotes reaction. With porous catalysts, reaction occurs at the gas–solid interfaces, both at the outside boundaries and within the pores themselves. Because the pores contain so much more area than the exterior surface, most of the reaction takes place within the particle. The reaction products then diffuse out of the pores and back into the main gas stream.

The process proceeds in the following steps:

- Diffusion of the reactants from the bulk phase to the catalyst
- Diffusion of the reactants to the active sites
- Adsorption of the reactants at the active sites
- Electron transfer process at the active sites
- Desorption of the reaction products at the active sites
- Diffusion of the reaction products to the catalyst surface
- Diffusion of the reaction products to the bulk phase

During VOC oxidation, heat is released in the reaction step. If the heat release is too extreme, damage to the catalyst through the process of sintering may occur. Sintering causes a decrease in surface area and reduces the effectiveness of the catalyst.

The activity of oxidation catalysts can be described with a light-off curve. This is an S-shaped graph of conversion efficiency vs. temperature. The light-off curve is divided into three regions: (1) kinetically limited, (2) light-off, and (3) mass transfer limited. A generic illustration is shown in Figure 9.1.

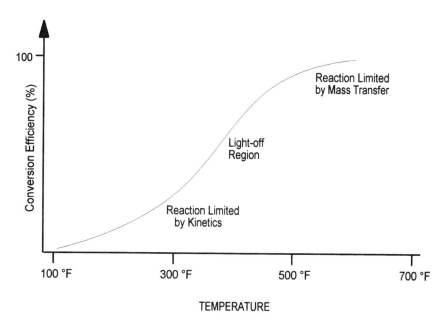

Figure 9.1   Generic catalyst light-off curve.

At low temperatures, the reaction is kinetically limited and any oxidation or conversion that occurs is dependent on interaction of the hydrocarbon and oxygen molecules at the catalyst surface. As the temperature increases, the reaction rate increases abruptly, due to the heat of reaction. This is the light-off region. It is characterized by a rapid increase in conversion efficiency over a narrow temperature range. The reaction then proceeds to the mass transfer limited region, where the reaction is only limited by the ability of the reactants to get to the catalyst sites. Each VOC has its own distinctive light-off curve.

The effectiveness of one catalyst compared to another can be compared using the light-off curve. Two points are identified on the curve. T50 (also known as the light-off temperature) is the temperature at which 50% of the VOC is oxidized. T90 is the temperature at which 90% of the VOC is oxidized. T90 is affected by the amount of unoxidized VOC on the surface of the catalyst. A light-off curve showing the effects of temperature on conversion for several VOC compounds is shown in Figure 9.2.

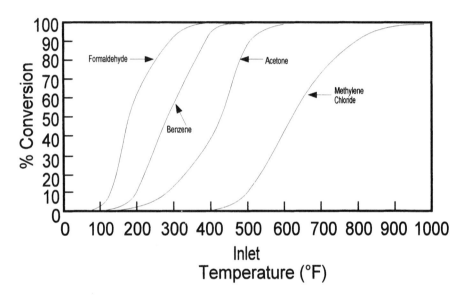

Figure 9.2   Oxidation of: formaldehyde, benzene, acetone, and methylene chloride. (Courtesy of Johnson Matthey.)

9.3 BASIC EQUIPMENT AND OPERATION

A schematic of a basic VOC catalytic oxidation system is shown in Figure 9.3. The waste gas is introduced into the mixing chamber, where it is heated to 400 to 800°F by the hot combustion products of an auxiliary fuel burner. The heated mixture then passes through a fixed bed of catalyst. Oxygen and VOC diffuse onto the catalyst surface and are adsorbed in the pores of the catalyst. The oxidation reaction takes place at these active sites. Reaction products are then desorbed and diffuse back into the gas.

Alternate schemes are shown in Figures 9.4 and 9.5. In Figure 9.4, the catalyst is part of a fluidized bed. Figure 9.5 shows another version of the fixed bed catalyst system. Here, the waste gas is preheated by the combustion products from the oxidation process. This is the most commonly used system and is similar to recuperative heat recovery with a standard thermal oxidation system.

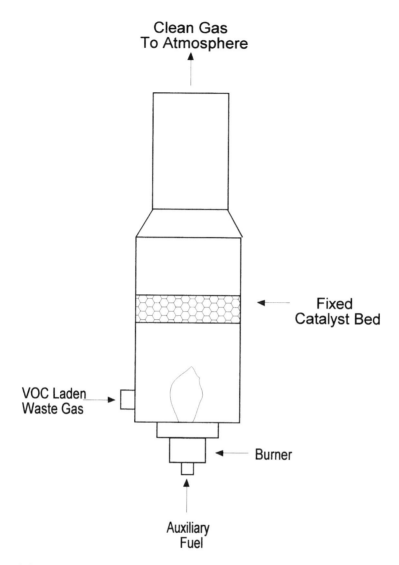

Figure 9.3 Basic catalytic oxidizer configuration.

9.4 GAS HOURLY SPACE VELOCITY (GHSV)

The amount of catalyst required in a VOC catalytic oxidation process is dependent on a parameter called the gas hourly space velocity or GHSV. The GHSV is defined as

$$\text{GHSV} = \frac{\text{Volumetric gas flow rate}}{\text{Catalyst volume}}$$

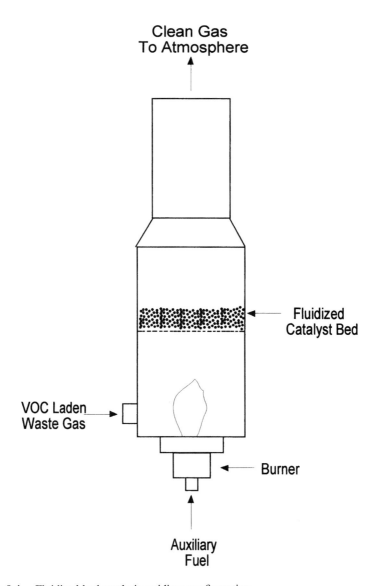

Figure 9.4   Fluidized bed catalytic oxidizer configuration.

where the units are
 GHSV (1/hr)
 Volumetric flow rate (scfh)
 Catalyst volume (ft³)

The higher the GHSV, the greater the reactivity of the catalyst and the less volume of catalyst required to achieve a specified VOC destruction efficiency. For mixtures of VOCs, the GHSV is determined for the most difficult VOC to oxidize. For specific

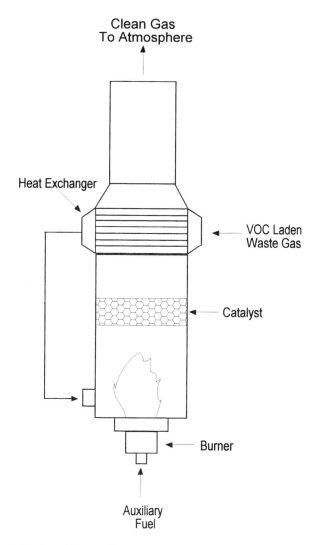

Figure 9.5   Catalytic oxidizer with waste gas preheat.

applications, GHSVs are sometimes determined by conducting pilot tests. Even though the flow rates between the pilot scale and full-scale units may be dramatically different, the performance should be the same if the GHSVs are identical.

The quantity of catalyst required for a given application can be optimized by judicious selection of the GHSV, substrate type, cell density, and catalyst formulation for a specific application. For example, to decrease the GHSV, the cell density could be increased. However, the pressure drop of gas flow through the catalyst substrate would increase. When pressure drop is an overriding consideration, the cell density could be decreased or the catalyst bed cross-sectional area increased while reducing the depth of the catalyst bed.

9.5   CATALYST DESIGN

The catalyst is composed of three primary components: (1) the active ingredient, (2) the substrate, and (3) the washcoat.

9.5.1   SUBSTRATE

The substrate is a thin solid material or small particle, either ceramic or metal, on which the catalyst material is deposited. The support itself may be catalytically active. It is also called a *carrier*.

The substrate may exist in various materials and forms. Most common in VOC oxidation applications is the monolithic honeycomb shown in Figure 9.6. This

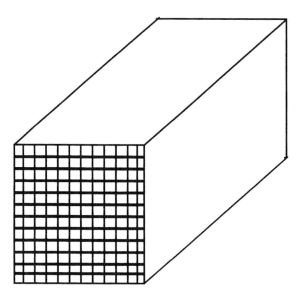

Figure 9.6   Honeycomb monolith catalyst substrate.

configuration provides the maximum surface area, while minimizing pressure drop. Both stainless steel and ceramic materials are used. Stainless steel substrates provide the best structural integrity and the lowest pressure drop. Stainless steel is especially applicable where frequent thermal cycling is expected. However, both stainless and ceramic materials have been used in a wide range of applications.

The substrate cell density can be varied for a given application. The pressure drop and available surface area increase as cell density increases. Cell densities of 4 to 400 cells per square inch (cpsi) have been used, although cell densities of approximately 200 are most common. The cell density can be varied to optimize the catalyst volume, VOC conversion efficiency, and pressure drop.

Ceramic pellets can also be used as catalyst substrates. However, they provide less geometric surface area per unit volume in comparison to monolithic supports.

They are suited to applications that contain known poisons or masking agents. The pellets act as a guard bed. Attrition of the catalyst surface over time abrades the catalyst, exposing fresh surface. At the same time, dust and particulate are removed. Additional catalyst pellets must be periodically added to account for the amount lost through attrition.

9.5.2 WASHCOAT

The washcoat is a material coated onto the catalyst substrate to provide a very high surface area support for the catalyst. This washcoat may contain oxides of aluminum, zirconium, transition metals, nickel, titanium, vanadium, cerium, or other metals.

Washcoat surface area is critical to the overall catalyst design. Washcoat formulations are selected, based on their ability to provide high surface area, thermal durability, poison tolerance, and reactivity. In addition to increasing the surface area of the catalyst support, the washcoat can be engineered to provide properties that will increase the catalyst life or enhance its performance. Components that can promote the VOC oxidation reaction can be added to the washcoat to enhance the catalyst's overall reactivity. The destruction of halogenated compounds is a strong function of the washcoat composition. Figure 9.7 is an illustration of the catalyst washcoat and substrate shown at the microscopic level.

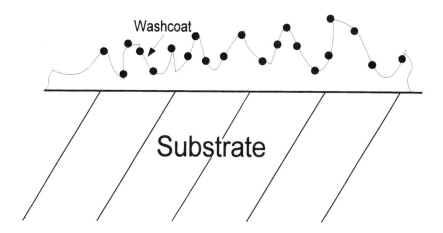

Figure 9.7   Catalyst components.

9.5.3 CATALYTIC MATERIALS

A number of materials are used to promote the VOC oxidation reaction. The most widely used are platinum group metals (a specifc grouping on the periodic table of the elements). Their prevalence is due to their high activity (and consequently reduced quantity of catalyst required), wide operating temperature range (300 to 1200°F), thermal durability, and resistance to poisons and deactivators. These precious metals are finely dispersed with fine particle sizes. This provides as many active reaction sites as possible to ensure high performance and long life.

Depending on the application, the catalyst may consist solely of platinum or palladium or a combination of platinum, palladium, or rhodium. The type of metals selected depends on the specific VOC or combination of VOCs that are to be treated. The catalyst chosen must also be selective to the desired oxidation reaction. Selectivity refers to a catalyst's ability to direct reactants to specific, desired products. For example, ethylene and oxygen on a platinum catalyst adsorb onto specific sites, combining at low temperatures and producing carbon dioxide and water vapor exclusively. However, if a copper/palladium catalyst is used with these same reactants, acetaldehyde is formed.

Base metal catalysts (typically oxides of chromium and manganese) have also been successfully used for VOC oxidation. They are typically supplied in pellet form and have higher pressure drops than catalysts on monolithic substrates. They also suffer from pellet attrition over time, although this can be an advantage in applications with waste gases containing particulate or dust. The costs of base metal catalysts are only 20 to 25% of those of platinum group metal catalysts.

The platinum group metal catalysts have high destruction efficiencies for aromatic compounds such as benzene, toluene, and xylenes. However, the temperature required for the oxidation of oxygenated VOCs (contains one or more oxygen atoms in its molecular structure) using platinum group metals is much higher as compared to metal oxide-based catalysts. Examples of oxygenated VOCs are alcohols, acetates, and ketones. These compounds are commonly found in the off-gas streams in the printing, coating, and chemical process industries.

9.6 OPERATION

In operation, the VOC-containing waste stream is injected just downstream of the burner, either without preheating or after preheating the waste stream. The burner raises the waste gas to the predetermined light-off temperature. As the VOC flows through the catalyst, oxidation occurs, raising the temperature. The pressure drop through the catalyst bed is typically 1 to 2 in. w.c.

If the waste stream contains little or no oxygen, oxygen is added by operating the burner with an appropriate quantity of excess air. Like a thermal oxidizer, a catalytic oxidizer must have a minimum of 2% oxygen in the combustion products to ensure complete reaction. Thus, the burner excess air rate is set to produce this concentration in the oxidation products. If the VOC is in a contaminated air stream, no further oxygen injection is required.

Since catalytic oxidizers operate at lower temperatures than thermal oxidizers, the internal lining is metal, usually stainless steel (Type 304 or 316), rather than refractory. Thermal insulation is provided around the internal lining and is encapsulated in a metal outer skin; aluminized steel, for example.

9.7   HALOGENS

As described earlier in this book, halogens comprise a group of chemical elements occupying a specific row in the periodic table of the elements. They are grouped together because they have similar properties. Waste gas streams containing halogenated species can present special problems for catalytic oxidation. Until the 1990s, catalytic oxidation was not used for VOC streams containing halogens because the halogen poisoned the catalyst. Catalyst formulations have since been developed which are not susceptible to poisoning by halogens. While the operating temperature is somewhat higher than a typical VOC, it is still much lower than required to achieve a higher degree of destruction with thermal oxidation. However, because of their lower operating temperature, much more Cl_2 is produced from treatment of chlorinated hydrocarbons compared to thermal oxidation. This is due to the temperature effect on the Cl_2/HCl equilibrium described in Chapter 4.

The presence of other VOCs can affect the light-off temperature of halogenated VOCs. For example, the light-off temperature of trichloroethylene is significantly reduced by the presence of ethylene. The design of catalyst systems for the destruction of halogenated species alone or in combination with other VOCs requires special attention to obtain the most suitable catalyst for the application.

9.8   CATALYTIC VS. THERMAL OXIDATION

The main advantage of catalytic oxidation vs. thermal oxidation is a much lower operating temperature. The following table compares the operating temperatures required for each technology for specific VOCs:

VOC	Catalytic	Thermal
Benzene	440	1550
Carbon tetrachloride	610	1650
Methyl ethyl ketone	600	1450
Hydrogen cyanide	480	1600

Temperature (°F) for 99% Destruction Efficiency

In addition to a lower operating temperature, the residence time required for a catalytic system is approximately 0.25 s, compared to the 0.75 to 1.0 s required for a thermal oxidizer to achieve 99% destruction efficiency.

Example 9.1: Compare the auxiliary fuel costs of a thermal oxidizer vs. a catalytic oxidizer for a 1000-scfm air stream contaminated with 100 ppmv of methyl ethyl ketone (MEK) at a temperature of 77°F. Assume an auxiliary fuel value of $3/MM Btu, 8400 hr/year operation, and no waste gas preheat on either unit.

Scfm of MEK = 1000 scfm $\times$ 100 $\times$ 10^{-6} = 0.10
MEK chemical formula = C_4H_8O
MEK molecular weight = 72.11

lb/hr of MEK = 0.1 scf/min × 60 min/Hr × 1 lb-mol/379 scf
   × 72.11 lb/lb-mol = 1.14
Heating value (Btu/lb − LHV) = 13,671
Heat released = 13,671 Btu/lb × 1.14 lb/hr = 15,607 Btu/Hr

Since the waste gas is a contaminated air stream, no additional air is needed for either system. This simplifies the calculations. To heat the entire waste gas stream to 600°F requires the following quantity of heat:

$$Q = m_{wg} \times Cp_{wg} \times \Delta T$$

m_{wg} = 1000 scf/min × 60 min/hr × lb-mol/379 scf × 28.85 lb/lb-mol = 4567 lb/hr

or

m_{wg} = 1000 scf/min × 60 min/hr × 0.076 lb/ft³ = 4560 lb/hr

Since this is a contaminated air stream, the mean heat capacity for air at 600°F (from Table 5.1) is 0.245 and 0.257 Btu/lb-°F at 1450°F.

Catalytic Oxidizer

$$Q = 4567 \text{ lb/hr} \times 0.245 \times (600°F − 77°F) = 585,193 \text{ Btu/hr}$$

Fuel cost = 0.59 MM Btu/hr × $3/MM Btu × 8400 hr/year = $14,868/Year

Thermal Oxidizer

$$Q = 4567 \text{ lb/hr} \times 0.257 \times (1450°F − 77°F) = 1,611, 516 \text{ Btu/hr}$$

Heat from MEK combustion = 15,607 Btu/hr

Net heat required = 1,611,516 − 15,607 = 1,595,909 Btu/hr = 1.6 MM Btu/hr

Fuel cost = 1.6 MM Btu/hr × $3/MM Btu × 8400 hr/year = $40,320/year

Thus, the annual fuel saving amounts to approximately $25,500 for the catalytic oxidizer when considering fuel usage alone.

Note one difference between the catalytic and thermal oxidizer. The heat generated from the VOC oxidation reaction contributes to the final operating temperature of the thermal oxidizer. The operating temperature specified for the catalytic oxidizer is at the entrance to the catalyst bed, where VOC oxidation has not yet occurred. The heat released from the VOC oxidation reaction contributes to a rise in the temperature of the combustion products at the exit of the catalyst bed. In this case, that temperature rise is

$$Q = m_{poc} \times Cp_{poc} \times \Delta T$$

$$15{,}607 \text{ Btu/hr} = 4567 \text{ lb/hr} \times 0.28 \text{ Btu/lb-}°F \times (T_f - 600)$$

$$T_f = 612°F$$

Temperature rise $= 12°F$

These calculations have ignored the increase in mass due to auxiliary fuel addition. The effect will be very minor in this case.

The process configuration for typical catalytic oxidizers includes a recuperative heat exchanger to preheat the waste gas before it enters the catalyst bed. In many cases, this preheat provides sufficient heat to avoid the need for any auxiliary fuel.

Example 9.2: The waste gas of Example 9.1 is preheated to 350°F in a recuperative heat exchanger before entering the combustion chamber of a catalytic oxidizer (see Figure 9.5). Determine the amount of auxiliary fuel required.

$$Q = 4567 \text{ lb/hr} \times 0.245 \times (600°F - 350°F) = 279{,}728 \text{ Btu/hr}$$

The annual fuel savings created by the installation of the recuperative heat exchanger is

$$\text{Annual fuel savings} = (0.59 - 0.28) \text{ MM Btu/hr} \times \$3/\text{MM Btu}$$
$$\times 8400 \text{ hr/year} = \$7{,}812/\text{year}$$

Justification for the use of the heat exchanger must be determined by a trade-off between its capital cost vs. annual fuel savings.

9.8.1 EMISSIONS COMPARISON

Another advantage of the catalytic oxidizer is NOx emissions. Since the operating temperature is lower and the amount of auxiliary fuel required is lower, the quantity of thermal NOx generated will also be less for the catalytic oxidizer. However, when treating VOC compounds that contain chemically bound nitrogen, both systems will generate NOx from the VOC. In this case, the thermal oxidizer has the advantage in that it can be staged to minimize chemical NOx formation.

9.9 WASTE GAS HEATING VALUE EFFECTS

Typically catalytic oxidizers are used to treat waste gases where their VOC concentration is less than 25% of the LEL. This generally is equivalent to a waste gas heating value of approximately 10 to 20 Btu/scf. When a VOC is oxidized over the catalyst, it releases this heat. Consequently, the temperature of the oxidation reaction products rises. If this temperature rise is too high, damage to the catalyst in the form of sintering may occur.

To prevent overheating, the temperature of the combustion products is monitored, and tempering air is added. This is shown diagrammatically in Figure 9.8. Ultimately, this increases auxiliary fuel consumption, since the light-off temperature required at the inlet to the catalyst remains the same. More fuel is needed to raise the temperature of the original waste gas stream plus the tempering air to the light-off temperature.

Example 9.3: A catalytic oxidizer is used to treat a 1000 scfm air stream at 77°F contaminated with methyl ethyl ketone (MEK). The catalytic oxidizer operates at a catalyst inlet temperature of 600°F. Normally, the MEK concentration is 1000 ppmv. However, during process upsets, it can spike to 7000 ppmv. (1) Determine the temperature of the combustion products during the upward spike in concentration assuming no tempering air is added.

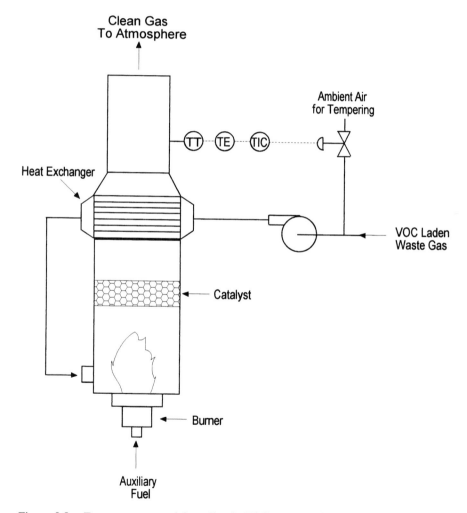

Figure 9.8   Temperature control for spikes in VOC concentration.

(2) Catalyst sintering will occur if the temperature exceeds 1200°F. Determine the quantity of tempering air needed to limit the temperature rise to this temperature.

Part 1

Scfm of MEK = 1000 scfm × 7000 × 10⁻⁶ = 7

MEK chemical formula = C_4H_8O

MEK molecular weight = 72.11

Lb/hr of MEK = 7 scf/min × 60 min/hr × 1 lb-mol/379 scf
    × 72.11 lb/lb-mol = 80

Heating value (Btu/lb – LHV) = 13,671

Heat released = 13,671 Btu/lb × 80 lb/hr = 1,093,680 Btu/hr

$$Q = m_{wg} \times Cp_{wg} \times (T_f - T_I)$$

T_I = 600°F (increased from the initial 77°F temperature by heat from burner)

m_{wg} = 1000 scf/min × 60 min/hr × lb-mol/379 scf × 28.85 lb/lb-mol = 4567 lb/hr

Since the final temperature is unknown, use a typical heat capacity value of 0.28 Btu/lb-°F.

$$1,093,680 \text{ Btu/hr} = 4567 \text{ lb/hr} \times 0.28 \times (T_f - 600°F)$$
$$T_f = 1455°F$$

To be completely accurate, the mean heat capacity of the combustion products at the final temperature should be used. Therefore, the products of combustion must be calculated, their corresponding mean heat capacity at 1455°F determined, the final temperature re-calculated using that heat capacity, and so on until the final temperature and heat capacity match.

Part 2

The temperature calculated in Part 1 is in excess of the safe operating limit by 255°F. This is equivalent to the following excess heat:

$$Q = m_{wg} \times Cp_{wg} \times \Delta T = 4567 \times 0.28 \times (255)$$

$$Q = 326,084 \text{ Btu/hr}$$

Assuming an initial air temperature of 60°F, the tempering air required is

$$Q = m_{air} \times Cp_{air} \times \Delta T$$

326,084 Btu/hr = m_{air} × 0.255 Btu/lb-°F × (1200 – 60) {Cp for air used}

$$m_{air} = 1122 \text{ lb/hr} = 246 \text{ scfm}$$

9.10   CATALYST DEACTIVATION

It would seem that the use of catalytic oxidation in comparison to thermal oxidation would be an obvious choice. However, there can be problems with catalytic oxidizers that are not encountered with a thermal oxidizer. Therefore, all factors must be considered before a decision is made as to which type of oxidizer is best suited to a particular application.

One of these factors is catalyst deactivation. This may result from chemical poisoning, masking, or thermal sintering. These processes occur slowly and are often referred to as "aging." Some of these processes are reversible and some are not. All catalyst undergo some deactivation over time. Sometimes an increase in the operating temperature can compensate for this deactivation. However, catalysts will age to a point beyond which their effectiveness diminishes such that they must be replaced.

9.10.1   POISONING

Poisoning is deactivation resulting from the chemical interference of materials in the process stream with the catalyst. This results when a contaminant reacts with the catalyst to form a new compound, rendering the site inactive. An example is when sulfur reacts with an alumina washcoat to create aluminum sulfate (Al_2SO_4). Poisoning may or may not be reversible. Catalyst poisons include heavy and base metals (e.g., lead, nickel, antimony, tin, arsenic, copper, mercury, chromium, and zinc), silicon, phosphorous, sulfur, dust, particulate matter, and some high molecular weight organic material.

9.10.2   MASKING

Here, the catalyst activity is reduced by gradual accumulation of noncombusted or inorganic solid material on the catalyst surface, preventing penetration of the gas stream to the catalyst sites. This problem typically arises from dust or dirt, metal oxides formed in the process, corrosion products from the ductwork, or organic char from operating at a temperature that was too low for complete oxidation. Masking can also occur on start-up of a system when the catalyst bed is cold. The VOCs can condense and plug the pores of the catalyst. When masking agents are known to be present, a regenerable catalyst should be selected. Another option is the use of a fluidized bed of catalyst. Here, the catalyst masking agent is abraded and elutriated, which automatically regenerates the catalyst sites.

9.10.3   SINTERING

Sintering is the agglomeration or densifying of the support material that contains active catalyst sites. This results when a catalyst is exposed to temperatures exceeding the design limit. Most catalysts cannot withstand temperatures above 1200°F for an extended period of time without damage.

Sintering reduces the total catalyst surface area exposed to the process gas. Above the recommended maximum operating temperature, the catalyst particles will

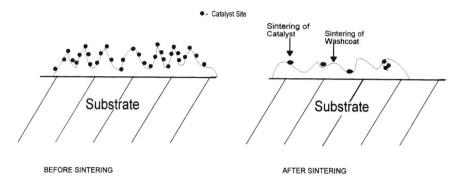

Figure 9.9   Catalyst and washcoat before and after sintering.

begin to agglomerate. At still higher temperatures, sintering of the washcoat will cause a collapse of the support material, resulting in a loss of surface area and catalyst activity. This is illustrated in Figure 9.9.

9.11   DEACTIVATION INDICATORS

The catalyst temperature rise and system pressure drop are indicators of performance. Normally, the catalyst supplier will provide expected temperature rise and pressure drop values for specific process conditions. Deviations from these expected operating conditions are an indicator of catalyst deactivation. For example, a decrease in the catalyst temperature rise may indicate a loss in catalyst activity. In some cases, this can be restored by elevating the operating temperature.

An increase in pressure drop across the catalyst bed may indicate that deposits are plugging the catalyst. Air lancing is sometimes sufficient to remove the deposits. Other times, chemical washing techniques must be employed. Many times filters are installed upstream of the catalytic oxidizer to prevent poisoning and masking agents from entering the process. While temperature rise and catalyst bed pressure drop can be an indicator of catalyst deactivation, they can also be used as an indicator that the catalyst activity is unchanged.

9.12   REGENERATION

In many cases, a catalyst can be regenerated using one of three techniques: (1) thermal, (2) physical, and (3) chemical cleaning.

Thermal Techniques

Thermal cleaning is used when the catalyst has been masked with organic compounds or char. To remove these masking agents, the catalyst bed temperature is elevated for a limited period of time sufficient to vaporize or oxidize the masking material. This is generally 100 to 200°F above the normal operating temperature. The increased temperature is obtained by operating the system burner at a higher firing rate.

Physical Cleaning

Physical cleaning uses mechanical means to remove dust and large particles. Vacuuming or blowing compressed air or water across the catalyst is used to dislodge any material present. This type of cleaning can be accomplished without removing the catalyst.

Chemical Treatment

This method involves removing the catalyst module from the unit and cleaning it in acid or alkaline cleaning solutions or a combination of both. This is the most common catalyst cleaning procedure. Chemical cleaning does not affect the catalyst composition; it merely removes masking agents from the surface of the catalyst. It can be done by the user in the field.

Sometimes a sample of the catalyst is removed for analysis. An activity test is performed to compare the catalysts' performance against a standard. If the test reveals deactivation, appropriate steps are taken to determine if chemical cleaning will be effective and the type of cleaning solution required. Figure 9.10 shows the normal sequence followed to determine the correct regeneration technique.

9.13   PERFORMANCE COMPARISON

Even after reactivation, the catalyst performance may not return to its original level. The light-off curve of Figure 9.11 compares the performance of a fresh catalyst, a contaminated catalyst, and a regenerated catalyst. With a regenerated catalyst, it may be necessary to operate at a higher temperature to offset the reduction in catalyst activity.

9.14   PILOT TESTING

When actual operating performance data are not available for a particular VOC or combination of VOCs in a waste stream or when the waste gas stream contains deactivators, pilot testing is warranted. On-site pilot testing offers a pragmatic, economical solution to determining the proper catalyst for a particular application. In addition, these tests conclusively demonstrate the effectiveness and durability of catalysts for that application. A pilot test is conducted using a portable unit to test a slipstream of the waste gas that requires treatment.

9.15   SUMMARY

In comparison to thermal oxidation, catalytic oxidation has lower auxiliary fuel costs for dilute waste gas streams. However, there are other factors that must be considered in the choice between a catalytic and thermal oxidizer. These include:

- System configuration (with or without recuperative heat recovery)
- Overall capital cost comparison
- Catalyst life and replacement or reactivation costs

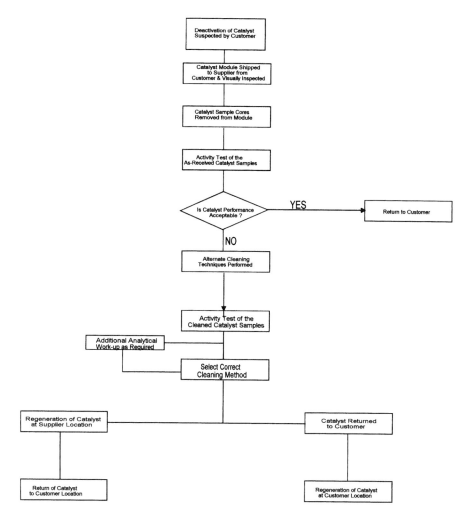

Figure 9.10 Logic diagram for catalyst regeneration.

- Catalyst VOC destruction efficiency with time
- Variability of the process VOC Concentration (are sudden spikes possible?)
- Potential presence of catalyst inhibitors in the waste gas stream
- Waste stream temperature and its variability

In general, a thermal oxidizer is far more tolerant of process upsets and variability in waste stream flow, composition, and temperature. Conversely, if the VOC concentration is dilute, the waste stream is free of particulates and inorganic species, and the waste stream relatively consistent, the potential fuel savings presented by catalytic oxidizers provides a strong incentive for their use. Since there are literally

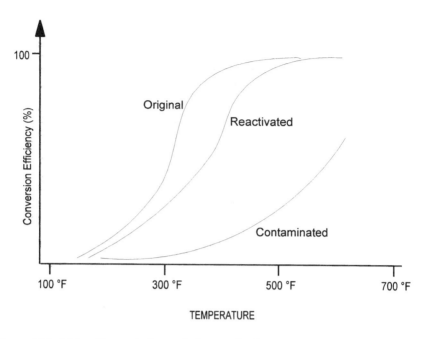

Figure 9.11   Light-off curve before and after reactivation.

thousands of both thermal and catalytic oxidizers in use, the need for both types of systems is obvious. Photographs of commercial catalytic oxidizer installations are shown in Figures 9.12 and 9.13.

Figure 9.12   Commercial catalytic oxidizer installation. (Courtesy of CSM Environmental Inc.)

Figure 9.13 Commercial catalytic oxidizer installation. (Courtesy of CSM Engineering Inc.)

A hybrid between a regenerative thermal oxidizer (RTO) and catalytic oxidizer emerged in the late 1990s. These regenerative catalytic oxidizers (RCO) are described Chapter 10.

10 Regenerative Systems

CONTENTS

The 1990 Clean Air Act Amendments directed the Environmental Protection Agency to establish Maximum Achievable Control Standards (MACT) for major sources of hazardous air pollutants (HAP). A major source is classified as one that emits more than 25 tons/year of a combination of HAPs or 10 tons/year of a single HAP. The pollutant species classified as HAPs are identified in Chapter 2. Prior to MACT standards, waste streams with low concentrations of VOCs were not regulated. However, with these new regulations, waste streams with low pollutant concentrations are regulated if the volume of the waste stream is large. For example, a 25,000-scfm waste stream containing only 10 ppmv of toluene is subject to MACT regulations.

Regenerative thermal oxidizers (RTO) are especially suited to applications with low VOC concentrations but high waste stream flow rates. This is due to their high thermal energy recovery, often as high as 97%. This means that VOCs in high volume waste streams can be thermally oxidized without the need for an excessive amount of auxiliary fuel. Thus, operating costs are minimized. In fact, many times the heat released from oxidation of the organic compounds present in the waste stream is sufficient to maintain the required operating temperature without the need for any auxiliary fuel.

The fuel costs savings of an RTO can be significant compared to no heat recovery whatsoever and even compared with recuperative heat recovery. This is demonstrated by the example shown in Table 10.1. In this example, the fuel costs for a RTO are less than 10% of those of a thermal oxidation system without heat recovery and approximately a fourth of the costs of a thermal oxidation system with recuperative (waste stream preheat) heat recovery.

TABLE 10.1
Fuel Cost Comparison for Various Modes of Thermal Oxidation Heat Recovery

	Annual Fuel Cost ($)
Thermal oxidizer without heat recovery	1,046,000
Thermal oxidizer with recuperative heat recovery (70% recovery)	345,000
Regenerative thermal oxidizer with 95% heat recovery	95,000

Assumptions: 25,000 scfm waste stream flow rate
Waste stream is contaminated air (no additional air added)
No heat contribution from VOCs
1400°F oxidizer temperature
Auxiliary fuel cost of $3/MM Btu
8400 hours/year operation

10.1   EVOLUTION OF THE RTO

Regenerative thermal oxidizers were first introduced to the marketplace in the early 1970s by REECO (Research-Cottrell). Even though the capital cost was much higher than the traditional thermal oxidizer, its fuel savings justified its use in a limited number of applications. Early units operated with 80 to 85% heat recovery efficiency (usually called "thermal energy recovery" — TER). The basic operation of an RTO consists of passing a hot gas stream over a heat sink material (usually a ceramic medium) in one direction and recovering that heat by passing a cold gas stream through that same heat sink material during an alternate cycle. This technology was borrowed from the glass industry, which had been employing a similar concept on glass furnaces since the early 1800s. In the glass industry, bricks were used as the heat sink material. The earliest RTOs used ceramic saddles as the heat sink material.

Saddles were originally developed for the chemical and petrochemical industries as a mass transfer packing for absorption and stripping towers. Saddles were an improvement over brick, due to their larger surface area (per unit mass) and lower mass. Also, the ceramic saddles were not susceptible to damage from any oxidation that occurred within the bed as the waste stream was heated.

As RTO designs evolved, the depth of ceramic saddles was increased. This increased the thermal energy recovery to ~95%. A rule of thumb in the industry was that an 8 to 9 foot (vertical) bed of 1" ceramic saddles was needed for a 95% TER. However, this also increased the pressure drop through the system. The corresponding electrical power supplied to fans to overcome this higher pressure drop became significant. Suppliers of these ceramic heat sink materials experimented with new designs to lower this pressure drop. One variation, called Typak, is shown in Figure 10.1.

Figure 10.1 Typak HSM RTO random packing. (Courtesy of Norton Chemical Process Products.)

Continued evolution of heat sink packings produced the structured packing. Both saddles and Typak are called random packings, because they are dumped into the RTO, forming irregular patterns. Gas flow through these beds is turbulent. In contrast, structured packings are manually installed in individual pieces and form regular, laminar flow patterns. A schematic of a structured packing is shown in Figure 10.2. Its similarity to the catalyst substrate shown in Figure 9.5 is more than coincidental. RTO equipment suppliers discovered that this traditional catalyst substrate could be used as a heat sink packing in RTOs. Structured packing not only decreased pressure drop through the system, but a much smaller volume of packing was required to achieve a given TER. On the other hand, the cost of structured packing was much greater per cubic foot than random packing. Consequently, in some applications, random packings may still be cost competitive.

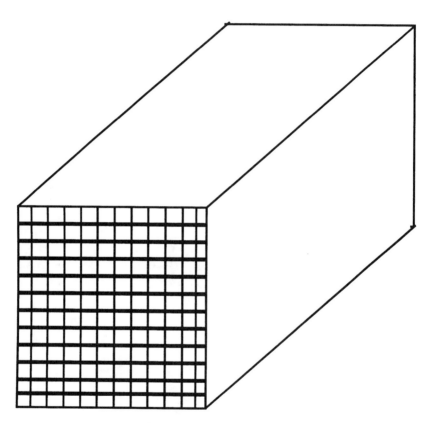

Figure 10.2 Structured packing for RTO applications.

Heat sink packing materials continue to evolve. Others will be discussed later in this chapter.

10.2 BASIC CONCEPT

The earlier RTO systems used three or more beds of ceramic heat transfer material to cyclically recover the heat from hot combustion gases. A schematic of the traditional three bed approach is shown in Figure 10.3. The beds are sometimes called regenerators. During normal operation, these ceramic beds contain stored heat from

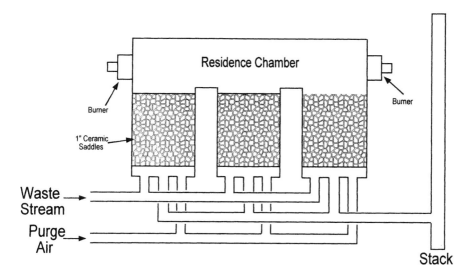

Figure 10.3 Traditional multiple-bed RTO design.

previous operating cycles. The contaminated waste gas stream enters one of the three beds (e.g., Bed #1). As it travels through the ceramic bed, heat is transferred from the ceramic medium to the gas stream. It exits the bed and enters the oxidation chamber at a temperature that approaches the final operating temperature of the oxidation chamber. A standard, gas-fired burner is then used to raise the temperature of the now preheated waste stream to the final operating temperature.

The hot combustion off-gases then exit the RTO through one of the remaining beds (e.g., Bed #3), transferring most of their heat to the ceramic heat transfer medium for recovery in a reverse cycle. During this reverse cycle, waste gas is diverted to enter through the third bed (Bed #3) and exits through Bed #2. Bed #2 was purged during the previous cycle. At the same time, fresh air is used to purge residual gases remaining in Bed #1 into the residence chamber for destruction of their VOCs. Cycles are then repeated, alternately cooling one bed, heating another, and purging the third.

This standard approach has been very effective in achieving high VOC destruction rates along with high heat recovery efficiencies. However, the use of multiple beds is expensive, requires a large plot area, and the resulting equipment is very heavy. In the early 1990s, designs using only two ceramic heat transfer beds emerged. A photograph of a commercial two-bed RTO is shown in Figure 10.4.

Generally, two-bed systems are very similar to multiple (three or more)-bed concepts, but without the purge cycle. Without a purge cycle, a portion of the waste stream at the end of a cycle remains untreated. Some fraction of the VOC-containing waste stream enters the heat sink bed but cannot enter the residence chamber before the cycle is reversed.

When the cycle is reversed, this untreated gas is discharged to the stack without treatment of its VOCs. The VOC destruction efficiency will typically

Figure 10.4 Two-chamber commercial RTO installation. (Courtesy of Trinity Air Technologies, Inc.)

exceed 99% during the extended portion of the cycle. However, a spike of VOCs in the stack discharge will occur when the cycle is reversed, thereby reducing the overall VOC destruction efficiency. With two bed systems, the VOC destruction efficiency is limited to approximately 98%. The actual value is dependent on the volumes of the regenerator beds and the concentration of VOCs in the waste stream.

A new type of RTO emerged in the late 1990s. This concept uses a single chamber of heat sink material. A rotating, indexing, waste stream flow diverter sequentially directs the waste stream into and out of specific segments of the chamber during a cycle. By applying a sealing device against the rotor surface at one end, the rotor is divided into two sections, inlet and outlet, through which the VOC-laden and clean waste streams flow into and out of the RTO chamber. Several manufacturers are now promoting this RTO design. A schematic of one such system is shown in Figure 10.5. An alternate is shown in Figure 10.6. Here, the heat sink beds themselves rotate to divert the waste stream to different segments.

10.3 THERMAL EFFICIENCY

One of the most important parameters in assessing a RTO's performance is its TER. It is typically defined for RTOs as follows:

$$\text{TER}(\%) = \frac{(T_{rc} - T_{savg})}{(T_{rc} - T_{wg})} \times 100$$

where

T_{rc} = residence chamber temperature (°F)
T_{savg} = average stack temperature over the cycle (°F)
T_{wg} = waste gas temperature entering RTO (°F)

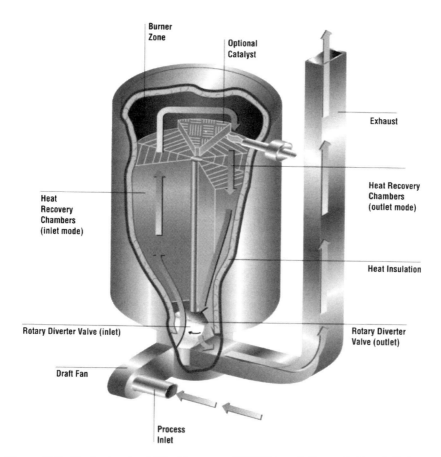

Figure 10.5 Single-chamber RTO. (Courtesy of REECO — A Research-Cottrel Co.)

This equation is applicable when the mass flow into the unit equals the mass flow out — that is, in cases where fuel and or air entering the residence chamber through a burner do not contribute significantly to the mass flow of the discharge stream. If there is a significant difference between the incoming waste gas flow and the outgoing stack flow, the TER must be calculated as follows:

$$\text{Thermal efficiency (\%)} = \frac{m_o(T_{rc} - T_{savg})}{m_i(T_{rc} - T_{wg})} \times 100$$

where
$$m_o \quad = \text{exhaust mass flow rate (lb/hr)}$$
$$m_i \quad = \text{incoming waste gas flow rate (lb/hr)}$$

These are the standard definitions of TER used in the RTO industry. In these equations, the "average" stack temperature must be used. In an RTO, this temperature varies over the duration of the cycle. Initially, the stack discharge temperature is

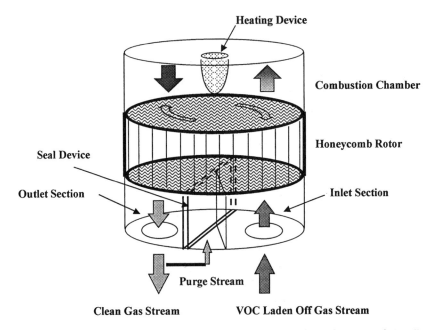

Figure 10.6 Single-chamber RTO with rotating heat sink packing. (Courtesy of Engelhard Corporation.)

relatively cold, but rises over the duration of the cycle. This occurs because the discharge end of the heat sink bed is relatively cold because relatively cold gas (waste stream) entered in the previous cycle. However, as combustion products exhaust through the bed, they begin to transfer their heat to the heat sink material, raising its temperature. Therefore, the entire heat sink bed is at different temperatures at various times during a cycle. This is illustrated by Figure 10.7, which compares the bed temperature profile at the beginning and end of a typical cycle.

> **Example 10.1:** A RTO operates with a residence chamber temperature of 1500°F and an average stack temperature of 200°F over the duration of the cycle. The temperature of the waste gas as it enters the RTO is 100°F. The VOC content of the waste stream releases sufficient heat during oxidation to maintain the required operating temperature without the addition of auxiliary fuel. Therefore, no additional mass is added to the exhaust gas. What is its TER efficiency?
>
> $$\text{TER} = \frac{(1500 - 200)}{(1500 - 100)} = 92.9\%$$

10.4   NUMBER OF HEAT SINK (REGENERATOR) BEDS

Early RTO designs used three or more heat sink beds. One was used to preheat the waste gas as it entered the RTO and the second to transfer heat to a heat sink bed in the exhaust cycle. The third bed was purged of untreated gas that entered during

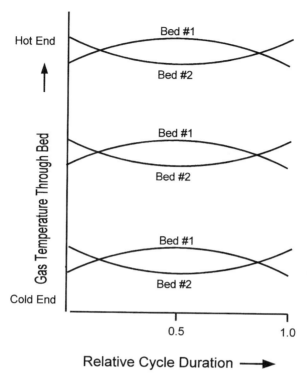

Figure 10.7 Typical heat sink bed temperature profile.

a previous cycle, thus preventing untreated gas from discharging to the stack when this bed was again used as the exhaust bed in the next cycle. While this arrangement is still in use, several techniques have been developed to capture and recycle the brief spike of VOCs in the exhaust gas that occurs when a cycle is reversed without the need for the third bed. In addition, many manufacturers have elected to use only two beds and accept a reduction in VOC destruction efficiency. Typically, the VOC destruction efficiency is limited to a maximum of 98% if a purge system is not included. However, in many cases, this is acceptable. The reduction in the number of beds required significantly reduces the capital cost of the system and the complexity of the control system.

10.5 PURGE SYSTEM

VOC destruction efficiencies in excess of 98% can be achieved in two ways: (1) operating with a relatively long cycle (and consequently reduced TER) or (2) by use of a purge system. Purge systems employ some type of arrangement to force untreated VOC remaining in the inlet bed of a previous cycle back into the residence chamber when a cycle is reversed. Traditionally this was done by using three heat sink beds. Purge air is forced through the bed used as the inlet bed in the previous cycle. Alternate techniques use an empty chamber or enlarged ductwork to collect these untreated gases for a short period (usually 1 to 2 s is adequate) and recycle these gases to the residence chamber during the next cycle. Purge systems are the subject of many RTO patents. While they do achieve their intended purpose of increasing the overall VOC destruction efficiency, they generally result in a small reduction in the TER because of an increased flow of relatively cold gas to the system. Purge systems also result in a somewhat more complex valving arrangement.

10.6   BED ORIENTATION

Most RTOs use vertical heat transfer beds. However, systems with horizontal beds do exist and function effectively. The heat transfer efficiency is nearly the same for beds of the same material and depth (length). Horizontal beds require a larger footprint. However, they do not require any support for the heat sink material. For large systems, the weight of the heat sink packing can be significant, and this weight must be supported to allow gases to flow underneath vertical bed systems. In both systems, distributing the incoming waste stream flow over the entire cross section of the heat sink packing is critical. If the flow distribution is unequal, the waste stream could bypass a portion of the heat sink bed and reduce TER. Methods used to avoid flow maldistribution are extended inlet lengths, turning vanes, a wire mesh that forces the flow to distribute evenly due to pressure drop, and a short layer of saddles, for example, under a bed of structured packing in a vertical unit.

10.7   THERMAL EFFICIENCY VS. CYCLE TIME

Specifying the thermal efficiency of an RTO without the corresponding cycle time can be misleading. Thermal efficiency is a function of design, operating variables, and type of heat sink packing. For a given design and packing type, the thermal efficiency decreases as the length of the cycle increases. This is illustrated in Figure 10.8. Therefore, TER values are meaningless without the corresponding cycle time.

High thermal efficiencies can be achieved with short cycle times. In fact, the thermal efficiency can exceed 95% for very short cycle times. The are two major disadvantages of short cycle times: (1) wear of valves and other equipment that actuate during each cycle and (2) lower overall VOC destruction efficiency for units without a purge system. The VOC destruction efficiency decreases because the short burst of VOC emissions when a cycle reverses becomes more frequent relative to the overall cycle time. Operators and maintenance personnel prefer longer cycle times because of reduced wear and maintenance.

10.8   HEAT SINK MATERIALS

10.8.1   RANDOM PACKING

All heat sink packings used in RTO applications are a ceramic of some type, typically composed mainly of silica (SiO_2) and alumina (Al_2O_3). Metals are not appropriate for this application because of their poor heat sink characteristics and because the

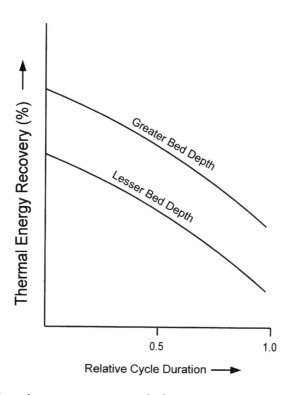

Figure 10.8 Thermal energy recovery vs. cycle time.

hot face of the heat sink bed is exposed to the very high oxidation chamber temperature. The first type of heat sink packing used in this application was the ceramic saddles traditionally used as tower packings in the chemical and petrochemical industries for mass transfer applications. Saddles are called a "random" packing because they are dumped into the bed chamber and form a random orientation and pattern. As stated previously, an early rule of thumb in the industry was that 8 to 9 ft of 1-in. ceramic saddles are required in an RTO to achieve a 95% thermal energy recovery efficiency with a 2-min process cycle time. Saddles are durable, but create a relatively high pressure drop through the system. They are also susceptible to chipping of the edges, which generates small pieces of ceramics within the packing void. Therefore, alternative heat sink media were developed.

Another type of random packing developed especially for RTO applications is shown in Figure 10.1. Called Typak-HSM, it has a 20 to 25% lower pressure drop than saddles for an equivalent volume of material and size. The thermal energy recovery is essentially the same. While the 1-in. size is the standard for comparison of both saddles and Typak-HSM, both can be purchased in 3/4 and 1 1/2 in. sizes. The 3/4-in. size is suitable for applications that require higher thermal efficiencies with shorter bed depths. The 1 1/2-in. size is suitable for applications that require increased capacity and throughput in revamped or new applications. The smaller sizes increase the thermal energy recovery efficiency for a given volume of material, but the pressure drop also increases. Conversely, with larger sizes, the pressure drop decreases but so does the TER efficiency.

10.8.2 STRUCTURED PACKINGS

Structured packings evolved for RTO applications in the early 1990s. These materials are characterized by a series of rectangular cells (sometimes called a honeycomb) forming a much larger block, as shown in Figure 10.2. These blocks typically have a 6 x 6-in. flow cross section with a 12-in. depth. However, longer blocks are available. These blocks are often called "monoliths." The flow through each individual cell is prevented from mixing with flow in adjacent cells because of the continuous nature of the cell walls. Therefore, in comparison to the turbulent flow through random packing, flow through structured packing is laminar at the normal operating conditions of a RTO.

Pressure drop varies with velocity through any conduit. In the turbulent flow regime, pressure drop is a function of the square of the velocity. That is, if the velocity doubles, the pressure drop is four times greater ($2^2 = 4$). However, in laminar flow, the pressure drop is a direct function of the velocity. That is, if the velocity doubles, the pressure drop doubles. Flow through random media is turbulent, while flow through structured monolithic media is laminar. Therefore, the pressure drop through structured monolithic media is usually much lower.

The structured monoliths are available for RTO applications in various cell densities. The cell density is the number of cells in the 6 x 6-in. flow cross section. Densities vary from 16 x 16 (256 cells) to 50 x 50 (2500 cells). They are usually characterized by the number of cells per square inch (cpsi). The greater the number of cells per square inch, the greater the surface area per unit volume. As with random packing, the greater the surface area per volume, the greater the heat transfer efficiency. Thus, the 50 x 50-cell monolith produces the greatest thermal energy recovery for a given volume of material. Physical characteristics of various monoliths are compared to Typak-HSM random packing in Table 10.2.

TABLE 10.2
Comparison of the Physical Characteristics of Random vs. Structured Packings

Property	Typak-HSM Random Packing			Structured Monolith Packing	
	3/4 in.	1 in.	1 1/2 in.	19 x 19 Cell	40 x 40 Cell
Bulk density (lb/ft³)	46–51	43–48	39–43	60	65
Surface area (ft²/ft³)	76	58	48	130	250
% Packing void	70	72	74	61	65

Another type of structured media was introduced to the marketplace in 1997. Called multi-layer media, MLM, it consists of a thin grid of interconnecting channels stacked one on another. Gas can flow both longitudinally and laterally in this packing. Each layer of packing is 1.5 mm thick and layers are stacked to form 12 x12 x 4-in.

Figure 10.9   MLM RTO heat sink packing. (Courtesy of Lantec Products, Inc.)

modules. A photograph of this type of packing is shown in Figure 10.9. Physical characteristics of this type of packing are compared with 1-in. saddles, 1-in. Typak, and 40 x 40-cell monolith in Table 10.3. The volume of this type of packing required to achieve a specified thermal energy recovery is less than that of the monolithic packing. However, at a given flow velocity, the pressure drop is approximately 50% higher than monolith, although still much lower than saddles. The cost of the MLM

TABLE 10.3
Comparison of Physical Characteristics of Packing Materials

	1-in. Typak	1-in. Saddles	40 x 40-Cell Monolith	MLM
Bulk density (lb/ft^3)	46	42	65	72
Surface area (ft^2/ft^3)	58	65	245	210
% Packing void	72	69	65	60

packing is lower than the monolith, so again a trade-off between capital and operating costs must be performed for a given application. In one commercial RTO retrofit from 1-in. saddles to the MLM packing, the capacity of the unit was increased by 30%, despite using half the volume of packing media, without any reduction in thermal energy recovery or reduction in cycle time.

In general, the performance advantages of structured media vs. random are due to their high surface area per unit volume. The higher the value of this property, the higher the thermal energy recovery. However, mass of media is also needed to maintain reasonable cycle times. Generally, systems with structured media can have shorter cycle times because the quantity of material used is much less than systems with random packings. The greater the mass of the media, the shorter the cycle time. A wide range of thermal energy recoveries and packed bed pressure drops are attainable with various heat sink media. The objective of a RTO design is to optimize the system in terms of operating costs and capital equipment costs.

Example 10.2: Both structured and random packing are being considered for a RTO design. The random packing produces an overall system pressure drop of 22 in. w.c., while the pressure drop with structured packing is only 10 in. w.c. The capacity of the system is 20,000 scfm. What is the difference in annual electrical energy cost if the fan on the RTO is 65% efficient (combined mechanical and motor) and the cost of electrical energy is $0.07/kw-hr?

$$\text{Fan horsepower (Hp)} = \text{flow (scfm)} \times \text{pressure drop ("w.c.)} \times 1.573 \times 10^{-4}/\text{efficiency}$$

$$\text{Kilowatts} = \text{horsepower} \times 0.745$$

Media Type	Fan Hp	Annual Electrical Cost ($)
Random	107	46,741
Structured	48	21,246

The structured packing not only saves operating costs, but a smaller, less expensive fan is required.

10.9   FLOW DIVERTER VALVES

With over 30 manufacturers selling RTOs, a variety of system configurations exist. One component(s) that varies widely is the flow diverter valve(s). These vary from two- and three-way poppet valves, to butterfly valve arrangements, to rotary valves, to other proprietary valve designs. Sometimes valves are linked to a common actuator, while other times valves operate independently. Features key to all of these valves are (1) rapid actuation, (2) low leakage, and (3) ability to operate over millions of cycles.

The diverter valve(s) on a RTO simultaneously direct waste gas flow to the heat sink beds and exhaust flow to the stack. While one valve port is the inlet, an adjacent port is the exhaust. If the valves actuate slowly, a small quantity of the inlet waste stream can bypass directly to the exhaust. When this occurs, untreated VOCs are exhausted to the stack and the overall VOC destruction efficiency is reduced.

The same principle applies to leakage. If the valve does not seal completely when closed, a small quantity of incoming waste gas can leak directly to the exhaust. Again, the overall destruction efficiency will be reduced. This highlights another advantage of structured packings. Because of their lower pressure drop, the pressure drop across the diverter valve is lower, reducing valve leakage.

Typically, a RTO valve will be actuated every 30 to 120 s (twice for each full cycle). In a 24 hr/d year-round operation, the valve will be subjected to several million cycles. Therefore, a very durable, low maintenance valve must be used in this application.

10.10   SINGLE-CHAMBER DESIGN

For many years, most RTO equipment vendors offered designs with at least two separate regenerator beds. However, in the late 1990s, several vendors developed single chamber designs. In this concept, a single, cylindrical chamber of heat sink packing is used. Waste stream flow is vertical. The chamber is divided into pie-shaped segments. A single rotary diverter valve controls flow to the inlet and exhaust of particular segments, one segment for preheating the waste stream and the second for transferring heat to the heat sink media. The rotary valve ratchets to a particular position in an indexing motion. This sequentially directs flow into and out of the proper heat sink packing. This type of unidirectional valve allows for fast rotation and optimum utilization of heat sink media. A photograph of one type of single chamber design is shown in Figure 10.5. In other versions of this concept, the entire heat sink medium chamber is rotated as illustrated in Figure 10.6. These designs are also reported to reduce the magnitude of pressure pulses that occur when cycles are reversed in the traditional dual/multibed concepts.

10.11   AUXILIARY FUEL INJECTION

Generally, one or more burners are installed on the residence chamber of an RTO. These are used for two purposes: (1) to bring the system up to temperature from a cold start and (2) to provide the remaining heat to raise the waste stream from the heat sink bed discharge temperature to the specified residence chamber operating temperature during steady-state operation. In some designs, auxiliary fuel is mixed with the incoming waste stream to provide the heat necessary to achieve the final temperature. This has two advantages: (1) additional mass is not added to the system as can occur when ambient air is used as combustion air in the burner and (2) nitrogen oxide (NOx) emissions are very low because of flameless oxidation. When a burner is used for this purpose, the waste stream itself is sometimes used as the source of combustion air. While this mitigates the effect of the additional heat load, this portion of the waste stream has not been preheated through the heat sink beds, and thus still increases the system heat load in comparison to in-bed auxiliary fuel injection.

In some systems, the heat release from the VOC itself is sufficient to sustain the specified operating temperature. Figure 10.10 shows the waste stream heat content needed for self-sustained oxidation at various operating temperatures and thermal energy recovery efficiencies.

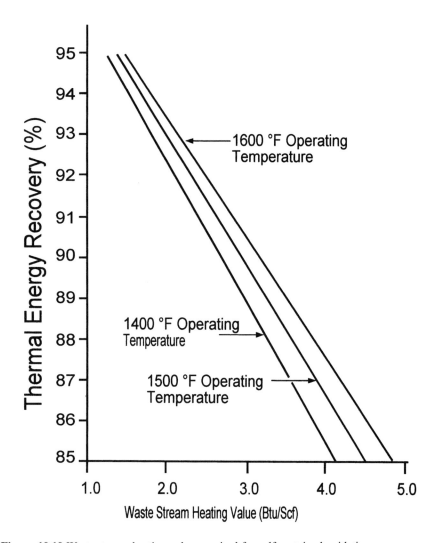

Figure 10.10 Waste stream heating value required for self-sustained oxidation.

10.12 POLLUTANT EMISSIONS

While the VOC destruction efficiency is a function of RTO design and operation, emissions of NOx and carbon monoxide (CO) are generally lower in comparison to the basic and recuperative thermal oxidizer designs. This is particularly true for

NOx. Because of their high heat recovery efficiencies, RTOs require little or no auxiliary fuel. Thermal NOx is generated primarily in a burner flame. With a reduced burner firing rate (or no burner firing whatsoever with in-bed auxiliary fuel addition), the overall NOx emissions are low, typically less than 10 ppmv.

The exception is when the waste stream contains nitrogen-bearing components. A large fraction of these components will be converted to NOx (chemically bound NOx), regardless of the method of auxiliary fuel addition. NOx emissions will be discussed in more detail in Chapter 11.

10.13   EFFECT OF WASTE STREAM COMPONENTS ON DESIGN AND OPERATION

There are several common components of VOC-containing industrial waste streams that require special consideration in the design and operation of a RTO. Following is a discussion of these components and design features to accommodate them.

10.13.1   INORGANIC PARTICULATE

In general, RTOs are intended for particulate-free streams. Over time, particulate can plug the heat sink beds. Sometimes particulate cannot be avoided because it is formed in the combustion process itself from constituents of the waste stream (e.g., organosilicates). An advantage of structured packing is its greater tolerance to particulate. Because of the straight, laminar flow pattern through the rectangular cells, they are less prone to plugging from particulate in comparison to random packing. However, some plugging can occur at the leading edge of each block. It is best to remove particulates from a waste stream upstream of the RTO. Systems used to perform this function are fabric filters, dry electrostatic precipitators (ESP), and wet electrostatic precipitators (ESP).

10.13.2   ORGANIC PARTICULATE AND BAKE-OUT CYCLE

Sometimes the particulate formed in a RTO is organic (combustible) in nature. They emanate from organic aerosols in the waste stream or from organic constituents of the waste stream that condense at the cold end of the heat sink beds. These are usually controlled using a "bake-out" cycle. During a bake-out cycle, the temperature of the exhaust stream is raised by depleting the bed of its heat content by continuing flow in one direction without reversing cycles or by firing a burner in the residence chamber and raising the temperature entering the heat sink bed. The system is held in this mode of operation until the exhaust temperature reaches approximately 900°F. Valves in the path of this exhaust must be selected to withstand the "bake-out" temperature or be cooled in some manner. An increase in the pressure drop across the system may indicate the need for bake-out.

10.13.3   HALOGENATED AND SULFONATED ORGANICS

If the waste stream contains halogenated compounds or compounds containing sulfur, acid gases will be produced during the oxidation reactions. The transient nature of RTO operation compounds the problem of acid corrosion. To combat this problem, hard, corrosion-resistant refractories are required in place of the usual ceramic fiber refractory. Sometimes the metal lining behind the refractory is coated to prevent acid attack or a membrane lining is applied. High alloy, corrosion-resistant metals are required for media supports, along with discharge valves and generally any cold-end metal components of the RTO. As with other types of thermal oxidizers, high temperatures are needed to favor the formation of HCl over Cl_2 and SO_2 over SO_3.

10.13.4   ALKALI ATTACK

Certain waste gas streams, particularly some of those generated in the wood products industry, contain alkali metals (e.g., sodium, potassium). These constituents have been found to chemically attack and degrade some ceramic heat sink media. Media with higher alumina contents have been found to be the most resistant to this type of attack. For example, traditional ceramic saddles have an alumina content of approximately 25%. Actual field experience shows that media with alumina contents of 70% are particularly resistant to alkali attack.

10.13.5   OXYGEN CONCENTRATION

RTOs are generally applied to VOC-contaminated air streams. However, they can successfully treat oxygen-depleted waste streams. However, the oxygen concentration must be at least 4 to 7% to achieve high VOC destruction rates. Alternatively, air can be mixed with an oxygen-depleted waste stream before it is injected into the RTO.

10.13.6   VOC LOADING

Regenerative thermal oxidizers are generally used in applications where the concentration of VOCs in the waste stream is less than 15% of the lower explosive limit (LEL). Because of the high heat recovery efficiencies of these systems, concentrations above this level could drive the temperature up and overheat the system. A typical scenario would be an upward spike in the VOC concentration of a waste stream during a process upset. This additional heat of combustion generated from oxidation of these VOC spikes increases the temperature in the residence chamber and consequently the stack discharge temperature. The temperature limitation of system components could be exceeded in this scenario.

One method to accommodate these VOC spikes is to include a hot bypass in the RTO design. This is illustrated in Figure 10.11. The bypass diverts combustion products directly to the stack, bypassing the heat sink bed. This bypass must be carefully designed. The materials of construction of the bypass duct must be resistant to high temperatures or, alternatively, refractory lined. A hot bypass damper must be included, and the stack must now be designed to tolerate these higher temperatures.

A second type of bypass arrangement is shown in Figure 10.12. Here, a cold bypass is used. Part or all of the waste stream is diverted around the regenerator bed. Therefore, without the preheat, the final temperature of the waste stream can be more easily controlled.

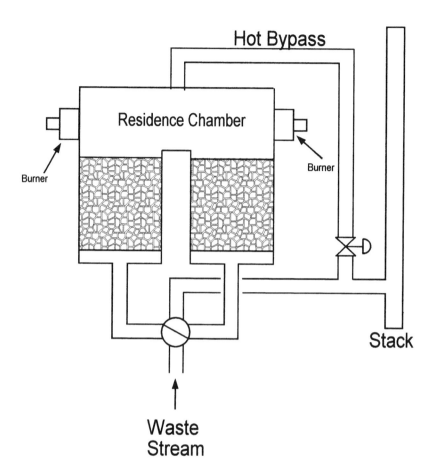

Figure 10.11 Hot bypass arrangement.

10.14   EXHAUST TEMPERATURE CONTROL

The temperatures of the combustion gases entering the stack vary over the period of the cycle. The thermal energy recovery is a direct function of the cycle duration. Usually, cycles are reversed, based on a timer. Alternately, the exhaust temperature can be monitored and the cycle reversed when this temperature reaches a prescribed value.

10.15   WASTE STREAM MOTIVE FORCE

Usually, the pressure of the waste stream as it discharges from the production process is insufficient to overcome the pressure drop through the RTO system. A fan, located either upstream (forced draft) or downstream (induced draft) of the RTO, is used to provide the motive force for flow through the unit. Under varying

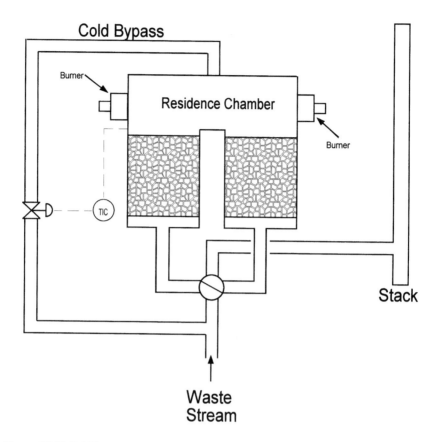

Figure 10.12 Cold bypass arrangement.

flow conditions, the fan pressure can be controlled with a damper or a variable frequency drive (VFD). The VFD option is particularly applicable to applications that contain multiple process sources. By varying the fan speed, the horsepower is adjusted to match the requirements of the process, and electrical energy consumption is minimized. The increased cost of the VFD must be compared to the electrical energy savings.

10.16   REGENERATIVE CATALYTIC OXIDIZERS

Regenerative catalytic oxidizers (RCO) emerged as a VOC control option in the mid to late 1990s. Just as catalytic oxidizers are an alternate to basic or recuperative thermal oxidation, regenerative catalytic oxidizers are an alternative to regenerative thermal oxidation (RTO). They are similar, except that a layer of catalyst is added to the top of each heat sink bed. The RCO concept is shown in Figure 10.13. Sometimes a thin layer of ceramic medium is added above the catalyst to protect it from radiant heat from the residence chamber. Again, compared to operation without

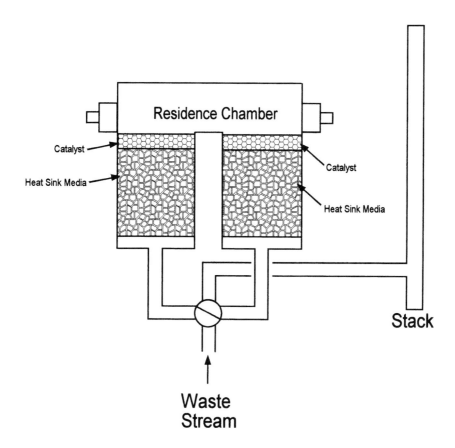

Figure 10.13 Regenerative catalytic oxidizer (RCO).

a catalyst, the operating temperature needed to achieve a given degree of VOC destruction is much lower. RCOs typically operate at temperatures of 600 to 800°F compared to 1400 to 1500°F for RTOs.

In RTOs, VOC destruction is essentially complete before the combustion gases exhaust through the discharge heat sink bed. However, with RCOs, the VOC oxidation reaction occurs both at the top of the incoming heat sink bed and the top of the outgoing heat sink bed. Combustion gases drawn off between the two beds would show inadequate VOC destruction. However, this illustrates another advantage of the RCO. The oxidation reaction occurs in the catalyst beds, not a residence chamber. Therefore, the residence chamber only serves as a conduit to direct gases from one bed to another and its size can be greatly reduced.

The RCO offers a reduction not only in auxiliary fuel costs but also in system pressure drop. Since the operating temperature required is reduced, less heat sink packing is required to preheat the waste stream before it enters the catalyst. Even with the catalyst addition, the overall system pressure drop of a RCO is less than a RTO.

Example 10.3: Compare the fuel and electrical energy savings of a RCO vs. a RTO in an application treating a 20,000 scfm waste stream at 100°F. The waste stream contains 40 lb/hr of a VOC with a heating value of 17,500 Btu/lb (LHV). To achieve the required VOC destruction efficiency, the RTO is operated at 1400°F and the RCO at 700°F. The pressure drop through the RTO system is 12 in. w.c. while only 8 in. w.c. through the RCO. The RTO has a 95% thermal energy recovery, while the RCO has a 92% TER. Assume 8400 hr/year operation, auxiliary fuel at $3.5/MM Btu, 65% overall fan efficiency, and electrical power at $0.07/kW-hr.

Using methods described previously,

	Auxiliary Fuel Required (MM Btu/hr)	Annual Fuel Cost ($)	Fan Horsepower Required (HP)	Annual Electrical Energy Costs ($)
RTO	0.81	23,814	58	25,407
RCO	0.39	11,466	39	17,084

10.16.1 CATALYST TYPE

Catalysts used in recuperative catalytic oxidation were described in Chapter 9. Those same types of catalysts are used in RCOs. Generally, the use of precious metal-based catalysts is more common. Base metal catalysts can emit products of incomplete combustion, such as aldehydes and carbon monoxide. The catalysts are coated onto both random and structured packing substrates.

While RCOs offer advantages in terms of auxiliary fuel and electrical power consumption, they also suffer from the same disadvantages described in Chapter 9. These include catalyst cost, catalyst replacement, aging, masking, and poisoning. Therefore, an economic comparison of a RCO to RTO must include life-cycle costs.

10.16.2 WASTE STREAM CONSTITUENTS

RCOs can generally treat waste streams with oxygen contents as low as 2 to 3%. Again, as described in Chapter 9, halogenated and sulfonated compounds in the waste stream require special attention to catalyst selection. In fact, this is one area where the lower operating temperature is a disadvantage. The much lower operating temperatures of RCOs favor the formation of chlorine gas (Cl_2) instead of HCl with chlorinated organics and SO_3 rather than SO_2 when sulfur compounds are present. These acid gases require special design consideration and materials selection to protect against corrosion. They are much more problematic if downstream scrubbing is required.

Spikes in the VOC loading also present special problems for the RCO. Whereas a bypass can be used in a RTO, a bypass in the RCO also bypasses one of the catalyst beds used for VOC oxidation. Thus, if a bypass must be used, the size of the catalyst beds must be adjusted accordingly.

10.17   RETROFIT OF RTO

The use of RTOs preceded the use of RCOs by many years. Some RTO installations have decided to retrofit their RTOs with RCOs. This can be accomplished by either adding catalyst on top of the existing heat sink beds (if room exists) or replacing part of the heat sink packing with catalyst. The catalyst is also a good heat sink material itself. If placed on top of existing heat sink media, the thermal energy recovery of the system can be enhanced.

11 Combustion NOx Control

CONTENTS

Thermal oxidation systems apply the principles of combustion to convert organic compounds to innocuous by-products. All combustion systems produce a small quantity of nitrogen oxides (NOx) as a by-product of the combustion reactions. Nitrogen oxides are one of the six chemical species classified as criteria pollutants under the National Ambient Air Quality Standards (NAAQS) of the Clean Air Act. NOx reduction is also one of the primary goals of Title I of the 1990 Clean Air Act Amendments.

NOx in combination with volatile organic compounds (VOC) present in the atmosphere can combine in the presence of sunlight to form ozone. Ozone has been found to be damaging to human health in concentrations as low as 0.1 ppmv. In 1992, the EPA reclassified areas of the U.S. for attainment of NAAQS standards. There were 185 areas classified as "non-attainment" for ozone. Both the federal government, through the 1990 Clean Air Act (CAA) Amendments, and individual states are stringently regulating NOx emissions from combustion sources. States must continue to tighten NOx limits until the NAAQS for ozone is attained.

There are two general methods used to reduce NOx emissions from combustion systems: combustion control and postcombustion control. The objective of combustion control is to prevent the formation of NOx during the combustion process. Postcombustion control deals with NOx reduction after it has formed. Both methods can be implemented concurrently to achieve the lowest overall stack NOx emissions. This chapter will discuss combustion control methods for NOx reduction. Postcombustion NOx control is discussed in Chapter 12.

11.1   CHARACTERIZING/CONVERTING NOx EMISSION LEVELS

Nitrogen oxides are actually a combination of two different chemical species, nitric oxide (NO) and nitrogen dioxide (NO_2). At normal combustion temperatures, the nitric oxide form predominates and is usually more than 95% of the total. However, as the combustion gases are emitted to the atmosphere through the stack, the nitric oxide reacts with the oxygen in the atmosphere as it cools to form nitrogen dioxide. Therefore, NOx emission calculations always assume that the NOx is in the nitrogen dioxide form (NO_2 molecular weight = 46)

NOx emissions are measured in units of parts per million by volume (ppmv). However, NOx emission limits set by environmental regulators are also written in terms of lb/hr (or tons/year) and pounds/million Btu of heat released by the combustion process (lb/MM Btu). The conversions between these different units are as follows:

$$\text{NOx (lb/MM Btu)} = \frac{\text{NOx (ppm, dry)} \times \text{dscf (POC)}/\text{MM Btu (HHV)} \times 46}{1,000,000 \times 379}$$

where

 dscf = dry standard cubic feet of products of combustion (POC)
 MM Btu (HHV) = heat release in MM Btu/hr (higher heating value)
 46 = molecular weight of nitrogen dioxide (NO_2)
 379 = scf/lb-mol of combustion products

$$\text{NOx (ppmv)} = \frac{\text{NOx (lb/hr)} \times 1,000,000}{\text{POC (lb-mol/hr)}} = \frac{\text{NOx (lb/hr)} \times 1,000,000}{\text{POC (scfm)} \times 60/379}$$

$$\text{NOx (ppmv, dry)} = \frac{\text{NOx (ppmv, as is)} \times 100}{(100 - \% \ H_2O \ \text{in POC})}$$

Generally, the instruments used to measure NOx emissions remove the water vapor before making the measurement. Thus, these results are on a dry basis. This must be taken into account when calculating emissions on a different basis. Often, the emission limits are specified at a certain oxygen concentration in the combustion products. The measured value is converted to the corrected value, using the following equation:

$$\text{NOx (ppmv, dry, corrected } O_2) = \frac{\text{NOx (ppmv, dry, measured } O_2) \times (21 - X)}{(21 - \text{Actual } \% \ O_2 \ \text{in POC})}$$

where X = base % O_2 being corrected to.

Example 11.1: A burner is firing natural gas (1013 Btu/scf) at a rate of 10 MM Btu/hr (HHV) with 25% excess air. NOx emissions were measured at a concentration of 75 ppmv (dry). What is the burner NOx generation rate in terms of lb/MM Btu of heat released? Correct the measured NOx value to 3% (dry).

By methods described in Chapter 5, the combustion products consist of the following:

Combustion Product	lb-mol/hr	Scfh	vol%
Carbon dioxide	26.05	9,900	7.75
Water vapor	52.10	19,740	15.50
Nitrogen	244.97	92,820	72.88
Oxygen	13.03	4,938	3.88
		127,380	

$$dscf = 9900 + 92820 + 4938 = 107,658$$
(as shown above but excluding water vapor)

The dry oxygen concentration is

$$\% \ O_2(dry) = \frac{4,938}{107,658} \times 100 = 4.59$$

$$NOx \ (lb/MM \ Btu) = \frac{75 \ ppmv \times 107,658 \ dscf \ (POC)/10 \ MM \ Btu \ (HHV) \times 46}{1,000,000 \times 379}$$

$$NOx \ (lb/MM \ Btu) = 0.098$$

$$NOx \ (ppmv, \ dry, \ 3\% \ O_2) = \frac{75 \ (ppmv, \ dry, \ 4.59\% \ O_2) \times (21 - 3)}{(21 - 4.59)} = 82$$

11.2   NOx FORMATION MECHANISMS

NOx is formed by one of the three mechanisms in a combustion process: thermal NOx, fuel or chemically bound NOx, and prompt NOx. Most NOx emissions from combustion processes are generated from thermal fixation of nitrogen in the combustion air. The generally accepted mechanism of thermal NOx formation is described by the Zeldovich equilibrium reactions shown below.

$$(1) \qquad N_2 + O^* \leftrightarrow NO + N^*$$

$$(2) \qquad O_2 + N^* \leftrightarrow NO + O^*$$

The N^* and O^* are radical species produced by the thermal dissociation of N_2 and O_2 at elevated temperatures. Reducing the peak flame temperature in a burner is a well-established method of reducing the NOx generation rate. The thermal NOx generation is also affected by the time at temperature. The longer the time, the greater the quantity of NOx generated.

Fuel or chemically bound NOx is generated from nitrogen compounds present in the waste or in the fuel itself. Generally, gaseous fuels such as natural gas or propane are free of nitrogen compounds. However, a significant amount of fuel-bound NOx can be generated from liquid fuels such as fuel oils, which can contain as much as 1% of nitrogen by weight. Of the 188 compounds listed as hazardous air pollutants (HAPs) under Title III of the Clean Air Act amendments, 42 contain nitrogen as part of their molecular structure.

Fuel or waste nitrogen compounds are only partially converted to the equivalent amount of NOx. The rate of conversion is much less than 1/1 in most cases. The exact conversion rate is a complex function of stoichiometry, temperature, and the specific nitrogen compound oxidized. However, for most compounds and conditions, the conversion rate is in the range of 20 to 70%. Generally, the lower the concentration of chemically bound nitrogen compounds present, the higher the percentage conversion to NOx and vice versa. Table 11.1 shows typical ranges. However, the true conversion rate is a function of the specific nitrogen-containing VOC compound and the exact combustion environment to which it is exposed.

TABLE 11.1
Chemically Bound Nitrogen Conversion to NOx

Fuel Nitrogen as% of Total Organic Compounds	% Nitrogen Conversion to NOx
1.0	30–35
0.8	23–36
0.7	34–38
0.6	36–40
0.5	39–44
0.4	42–47
0.3	46–51
0.2	51–57
0.1	58–65
0.05	68–72

A lesser-known type of NOx formation is termed "prompt NOx." Here, hydrocarbon radicals (CH, CH_2, etc.) formed from fuel fragmentation react with nitrogen in the combustion air to form a hydrogen cyanide (HCN) intermediate. The HCN then reacts with oxygen and nitrogen in the combustion air to form nitrogen oxides as shown below.

$$CH^* + N_2 \leftrightarrow HCN + N^*$$
$$HCN + OH^* \leftrightarrow CN^* + H_2O$$
$$CN^* + O_2 \leftrightarrow NO + CO$$

The formation of prompt NOx is proportional to the number of carbon atoms present in the fuel and has a weak temperature dependence and a short lifetime. Prompt NOx is only significant in fuel-rich flames that inherently produce low NOx levels. Thus, prompt NOx is usually a minor contributor to overall NOx emissions.

11.3 THERMAL NOx EQUILIBRIUM/KINETICS

Chemical equilibrium dictates the ultimate reaction products, given enough time for those reactions to reach equilibrium. In this case, chemical equilibrium predicts that the thermal NOx generation rate will increase exponentially with temperature. The reaction and rate equation are

$$N_2 + O_2 \leftrightarrow 2\ NO$$

$$K_{eq} = \frac{(NO)^2}{(N_2) \times (O_2)}$$

where

(NO), (N_2), (O_2) = mole fractions of nitric oxide, nitrogen, and oxygen, respectively,

K_{eq} = equilibrium constant

The equilibrium constant is a function of temperature as follows:

$$K_{eq} = 21.9 \times e^{-43,400/RT}$$

where

R = universal gas constant (1.988 cal/gr-mole-K)
T = absolute temperature (Kelvin)

Depending on gas concentrations, increasing the temperature from 2000 to 2600°F can increase the equilibrium NOx emission by 50 times. Also, note that the equilibrium equation contains oxygen concentration in the denominator. Although not as pronounced as the effect of temperature, higher oxygen concentrations in the combustion products also favor increased thermal NOx generation. In general, thermal NOx is an exponential function of temperature and varies with the square root of oxygen concentration.

While chemical equilibrium predictions match the trends observed in actual combustion systems, the values predicted usually overestimate observed NOx emissions. One reason is that NOx emissions are a function of reaction kinetics rather than chemical equilibrium. Residence time for reaction is a primary factor for reaction kinetics. Also, the flame is a heterogeneous mixture of different gas compositions at different temperatures throughout the flame envelope. Thus, a single temperature cannot represent the true conditions in a flame. The surrounding environment also impacts the NOx emissions. For example, NOx emissions from burners firing directly into boilers tend to be lower, because the boiler tubes extract radiant

TABLE 11.2
NOx Emission Factors for Natural Gas Combustion Systems

System Utility/Large Industrial Boilers	NOx (lb/10^6 ft^3 of natural gas)
Uncontrolled	550
Controlled — low NOx burners	81
Controlled — flue gas recirculation	53
Small industrial boilers	
Uncontrolled	140
Controlled — low NOx burners	81
Controlled — flue gas recirculation	30
Commercial boilers	
Uncontrolled	100
Controlled — low NOx burners	17
Controlled — flue gas recirculation	36

Source: Compilation of Air Pollutant Emission Factors, AP-42, 5 ed., Environmental Protection Agency, January 1995.

heat from the flame and lower its temperature. The same burner firing into a thermal oxidizer would produce higher NOx emissions.

Generally, NOx emissions are determined from testing burners and combustion systems under a variety of operating conditions. Estimates of NOx emissions from generic sources are available from "Compilation of Air Pollutant Emission Factors," developed by the Environmental Protection Agency. Better known as AP-42, these factors represent a value that attempts to relate the quantity of a pollutant released to the atmosphere with an activity associated with the release of that pollutant.

11.4 PARAMETRIC EFFECT

While NOx emissions are quoted in various sources, these data must be qualified by the operating conditions and equipment in which they were generated. These factors can have a profound effect on the values generated. In fact, it is the control of these factors which is used primarily to reduce NOx emissions. Those factors are discussed below.

11.4.1 TEMPERATURE

NOx emissions are affected by both the flame temperature and the bulk gas temperature after the combustion reactions are complete. Again, the higher the temperature, the higher the NOx emissions. This is shown in Figure 11.1. Increasing the thermal oxidizer temperature from 1600 to 2400°F doubles the NOx generation. These values are approximate, since they are very much equipment dependent and dependent on the method of injecting the waste stream into the combustion chamber. The effect of temperature is not as pronounced with low NOx burners. Rich waste streams tend

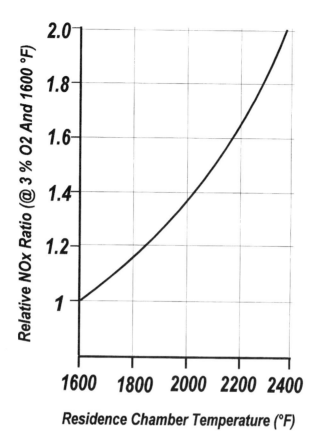

Figure 11.1 Effect of residence chamber temperature on NOx emissions.

to increase the temperature of a thermal oxidizer. NOx emissions are another reason to limit this temperature rise.

11.4.2   COMBUSTION AIR PREHEAT

As described in Chapter 8, the combustion air can be preheated to reduce the auxiliary fuel required in a thermal oxidation system. However, this has one drawback, increased NOx emissions. This is illustrated in Figure 11.2. Again, the effect is less pronounced if a low NOx burner is used.

There are two exceptions to this effect: staged combustion systems and regenerative thermal oxidizers (RTOs). In staged combustion systems, which are comprised of two separate zones characterized by an air deficiency in one and excess air in the other, preheating the combustion air does not significantly affect NOx emissions. Also, in RTOs which operate without the peak temperatures encountered in a burner, preheating the waste stream as it travels through the heat sink bed does not increase NOx emissions. In fact, low NOx emissions are a characteristic of RTO systems.

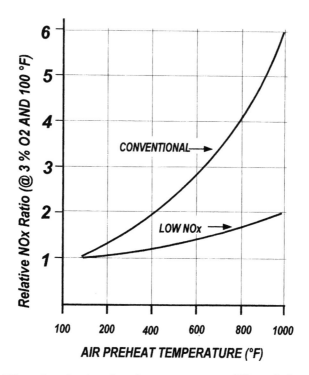

Figure 11.2   Effect of combustion air preheat temperature on NOx emissions.

11.4.3   EXCESS AIR

The NOx equilibrium equation shows that NOx is also a function of the oxygen concentration of the combustion products. Therefore, operating at low excess air levels (lower oxygen concentrations in the combustion products) can reduce thermal NOx generation. The standard oxygen content used in design by most thermal oxidizer manufacturers is 3% oxygen in the combustion products. While there is little difference in VOC destruction and CO emissions if this concentration is lowered to 2%, further reductions can result in reduced VOC destruction efficiency. Also, CO emissions begin to increase below this level and can be particularly high if the oxygen concentration is reduced to below 1%.

Conversely, if the excess air level is very high, NOx emissions will be reduced. This is due to the cooling effect of the sensible heat load of this excess (unreacting) air stream. Again, thermal NOx is a strong function of temperature. Reducing the flame temperature of a burner is one method of reducing NOx emissions. However, as with very low excess air levels, very high levels of excess air can increase CO emissions from most burners.

Figure 11.3 shows the relationship between excess air and NOx levels. The curve is shown to illustrate the effect of excess air on NOx emissions in general. The exact oxygen concentration at which the peak occurs in this curve is dependent on a

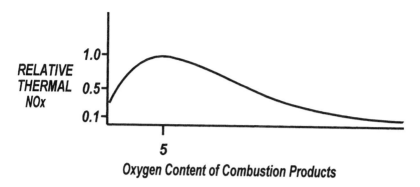

Figure 11.3   Thermal NOx vs. excess air.

particular burner and system into which it is fired. Thus, the peak will occur at different concentrations for different systems.

11.4.4   OXYGEN CONTENT OF COMBUSTION AIR

In general, ambient air or a contaminated air waste stream is used as the oxygen source for combustion in a thermal oxidizer. However, a reduced oxygen stream can be used as long as its oxygen content is sufficient to maintain a stable flame, as was shown in Figure 6.4. The lower the oxygen content, the lower the NOx emissions. This has the same effect as flue gas recirculation (FGR), one of thermal NOx control techniques which will be described later. The oxygen concentration in the flame envelope is reduced. As stated previously, thermal NOx emissions are proportional to the square root of the oxygen concentration. As a generalization, reducing the oxygen content of the combustion air from 21 to 16% will reduce NOx emissions by approximately half.

11.5   FUEL TYPE EFFECTS

The fuel used as auxiliary fuel in a thermal oxidation system can also affect NOx emissions. This effect is due to differences in adiabatic flame temperature with gaseous fuels and fuel nitrogen content with liquid fuels.

The higher the adiabatic temperature of a flame, the higher the NOx emissions. Different gases produce different adiabatic flame temperatures. Hydrogen, for one, has a particularly high adiabatic flame temperature and will increase NOx emissions as illustrated in Figure 11.4. The adiabatic flame temperatures of some pure gaseous fuels and common industrial fuel mixtures are shown in Table 11.3.

Fuel oils are the most common liquid fuels used as auxiliary fuel in thermal oxidation systems. Fuels oils fall into numeric classifications from No. 1 to 6 depending on their properties. The lower the number, the less dense the oil. Fuel oils with lower numbers also contain fewer impurities such as sulfur and ash. No. 2 and 6 fuel oils are

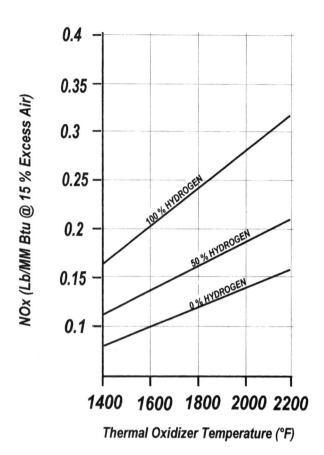

Figure 11.4 Effect of fuel hydrogen content on thermal NOx generation.

the most commonly used in thermal oxidizer applications. No. 2 is relatively pure, with a sulfur content typically in the range of 0.4 to 0.7% and 0 to 0.1% of nitrogen. While No. 2 fuel oil will produce slightly higher NOx emissions in comparison to gaseous fuels, the difference is usually on the order of 20 to 25%. In contrast, No. 6 fuel oil can contain high levels of sulfur (up to 3%) and nitrogen (up to 0.3%). This oil will not only convert the sulfur to sulfur dioxide in the combustion process but can also produce very high NOx emissions, depending on the exact concentration of nitrogen in the oil. Approximately 45% of the fuel oil nitrogen content will be converted to NOx.

11.6 NOx PREDICTION

NOx emission trends with equipment design and operating conditions are well known. However, a generalized prediction of the exact magnitude of NOx emissions is difficult since it is dependent on a number of factors, including thermal oxidizer design, burner design, method of waste stream injection, mixing, and the proportion of heat from the VOC vs. auxiliary fuel.

TABLE 11.3
Adiabatic Flame Temperatures of Common Fuels and Fuel Mixtures (Stoichiometric Air)

Fuel/Fuel Mixture	Adiabatic Flame Temperature (°F)
Carbon monoxide	4311
Hydrogen	4145
Natural gas (methane)	3773
Propane	3905
Acetylene	4250
Ammonia	3395
Butane	3780
Propylene	3830
Methanol	3610
Hydrogen cyanide	4250
Refinery gas (591 Btu/scf – LHV)	3832
Low Btu coal gas (148 Btu/scf – LHV)	3137
Blast furnace gas (LHV = 92 Btu/scf)	2582
(25.6% CO, 12.6% CO_2, 59.8% N_2, 1.3% H_2, 0.7% CH_4)	
Landfill gas (LHV = 447 Btu/scf)	3420
(41% CO_2, 6.6% N_2, 1.6% H_2O, 1.8% O_2, 49% CH_4)	
Coke oven gas (LHV = 471 Btu/scf)	3929
(7.4% CO, 2.0% CO_2, 5.6% N_2, 54% H_2, 0.4% O_2, 28% CH_4, 2.6% C_2H_6)	

If the waste stream is rich (high Btu content) and is injected through the burner, it will generate higher NOx emissions than a lean waste gas (low Btu content) injected downstream of the burner. This also illustrates one misconception of low NOx burners. If the auxiliary fuel-firing rate is low compared to the waste stream heat release, and the waste stream is not injected through the burner, then a low-NOx burner will have little effect on the overall NOx emissions level. That is, unless the burner is supplying a substantial proportion of heat to the system, a low-NOx burner may not be appropriate for NOx emissions control. In many cases, waste gas streams with relatively low heating values (25 to 50 Btu/scf) can provide most of the heat energy required to achieve the final operating temperature. In these cases, low NOx burners are of little help in reducing NOx emissions.

11.7 THERMAL NOx REDUCTION STRATEGY

Applying combustion control techniques can reduce NOx emissions produced by thermal generation. A number of techniques are available to the combustion system designer, but they all revolve around three general principles: (1) lower temperatures, (2) lower oxygen concentrations in the flame envelope, and (3) reduced residence time at peak temperatures. A discussion of specific methods applied to reduce thermal NOx generation follows.

11.8   LOW NOx BURNERS

If a substantial percentage of the heat released in the thermal oxidation system is released through the burner, then a low NOx burner can be used to reduce NOx emissions. There are several varieties of low NOx burners which reduce NOx through (1) recirculation of relatively cold combustion products back into the flame envelope; (2) operation at high excess air rates to reduce peak flame temperatures; (3) staged combustion air injection; and (4) staged auxiliary fuel injection.

Relative to the combustion flame of a burner, postcombustion gases are cooler. Recirculation of these relatively cool gases back into the flame envelope characterizes one category of low NOx burner. This is typically achieved by a bluff body, swirl vortex, baffle geometry, high velocities, or a toroidal ring. The geometry of the burner is designed to control this recirculation to provide optimum conditions in specific zones of the flame. The more efficiently this is achieved, the lower the NOx emissions.

In other types of low NOx burner, high excess air rates (60 to 100%) are used to reduce the peak flame temperature. The auxiliary fuel and air are typically premixed in a plenum upstream of the burner or in the burner ports themselves. In any event, the air and fuel must be intimately mixed before being ignited. This type of low NOx burner is sometimes called "flameless," since little visible flame is evident at the high excess air rates. The lack of a clearly definable flame is an indication of low temperature combustion.

A third type of low NOx burners splits the combustion air to form two zones within the burner. Approximately 60 to 70% of the theoretically required air is injected in the first zone in which all of the fuel is partially combusted. This fuel-rich combustion reduces NOx because of the reduced availability of oxygen to react. The air required to oxidize the residual carbon monoxide and hydrogen produced in the first stage is injected in the second zone of the burner. Peak flame temperatures are avoided in both stages, again limiting NOx emissions. These burners are also very effective in reducing NOx from high nitrogen-containing auxiliary fuels such as fuel oils.

The final low NOx burner design technique is called "fuel staging." A portion of the fuel is injected into the combustion air and burned in a very lean (high excess air) combustion zone. The NOx generation rate is low due to the low flame temperatures. The remainder of the fuel is injected downstream of the primary zone. NOx emissions from this secondary combustion zone are suppressed by the inerts emanating from the primary combustion zone. These inerts reduce the peak flame temperature and reduce the oxygen concentration.

Generally, an air-staged low NOx burner can reduce NOx emissions by approximately 30 to 50% compared to a conventional burner. A fuel-staged low NOx burner can lower NOx emissions to one third of those generated by a conventional burner. Some high excess air burners can achieve even better results. However, the high level of excess air can be a drawback in certain situations if that air is not needed and the additional heat load requires additional auxiliary fuel firing. Also, air and fuel staged low NOx burners tend to have longer and narrower flames than standard burners. This can impact the size of a thermal oxidizer.

11.9 VITIATED AIR

Vitiated air is air with reduced oxygen content. Firing a combustion system using vitiated air as the source of combustion air will also reduce NOx emissions. In fact, this air can be a waste stream with a reduced oxygen content or can be produced by blending a waste stream with ambient air as shown in Figure 11.5. When blending a waste stream with combustion air, the organic concentration of the mixture must be checked to ensure that it is not within the explosive limit or must include other features (discussed in Chapter 14) in the design to prevent flashback.

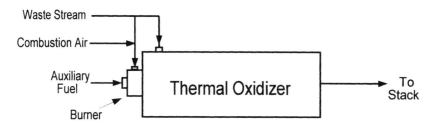

Figure 11.5 Burner air vitiation using the waste stream.

11.10 FLUE GAS RECIRCULATION (FGR)

Another common method used to reduce thermal NOx generation in a combustion system is called flue gas recirculation (FGR). With this technique, a portion of the products of combustion are recycled and mixed with the combustion air. It is commonly applied with recuperative heat recovery where relatively cold products of combustion are available. A fan is usually used to recycle the combustion products. If they were not first cooled in the heat recovery device, their temperature would be too high for a fan. This technique is shown schematically in Figure 11.6. It is very similar in concept to the air vitiation previously described.

With FGR, NOx formation is suppressed via two mechanisms. First, the peak flame temperature is reduced due to the larger mass of gas heated (sensible heat

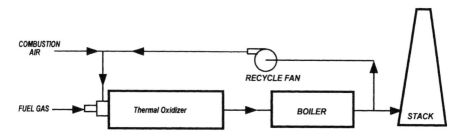

Figure 11.6 Flue gas recirculation.

load). Second, the oxygen concentration in the flame envelope is lowered. Lowering the peak flame temperature has the dominant effect on NOx reduction.

Flue gas can be recirculated to comprise up to 25 to 30% of the flue gas/air mixture. The quantity of flue gas recirculated is only limited by the burner stability at the lower oxygen content. NOx reductions can vary from 60 to 80% at high recirculation rates.

11.11   FUEL-INDUCED RECIRCULATION (FIR)

Fuel-induced recirculation (FIR) is similar to flue gas recirculation, except that the flue gas is mixed with auxiliary fuel upstream of the burner instead of combustion air. By diluting the fuel prior to combustion, the volatility of the mixture is reduced. This then reduces the concentration of hydrocarbon radicals that produce prompt NOx. FIR also reduces thermal NOx by acting as a sensible heat load in the same manner as FGR. Thus, FIR effects both thermal and prompt NOx formation.

11.12   WATER/STEAM INJECTION

Water can be injected, usually through a central liquid gun, through the burner flame to reduce NOx emissions. NOx reductions of up to 70% have been reported using this technique. It can have potential drawbacks, however. One is a possible effect on burner stability and the potential for formation of aldehydes and carbon monoxide in the reduced temperature flame. Second, if heat recovery is used downstream of the thermal oxidizer, the latent heat of the water evaporation cannot be recovered and thus the energy recovery is reduced.

Steam injection has a less pronounced effect on burner stability. Steam can be mixed with the auxiliary fuel or the combustion air. Adding steam to the fuel produces greater NOx reductions, but is not always an option in retrofitting existing equipment. In contrast to water injection, a portion of the sensible heat of the steam can be recovered in downstream recuperative heat recovery equipment. At low steam rates, NOx reduction is directly proportional to the quantity of steam injected. However, a point is reached where the response diminishes and with even higher rates the flame can become unstable. Maximum NOx reductions with steam injection are typically in the 20 to 30% range.

11.13   AIR/FUEL STAGING

Air and fuel staging have been described previously as they apply to low NOx burner designs. However, these same techniques can be used in the overall thermal oxidation system design. Air staging, as applied to the thermal oxidizer design, is usually applied to waste streams containing chemically bound nitrogen. This technique will be discussed in more detail later in this chapter.

Fuel staging can be applied in several ways. First, the auxiliary fuel can be staged. Usually, this fuel is mixed with high rates of air or waste gas to prevent flame formation at the injection point that could produce high NOx emissions. One auxiliary fuel staging concept is illustrated in Figure 11.7.

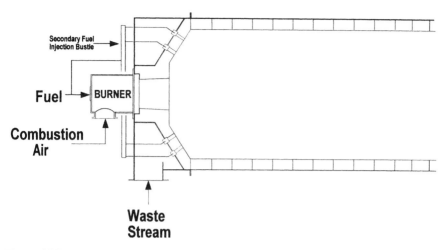

Figure 11.7 Fuel staging with natural gas for NOx reduction.

In a second concept, the waste stream itself can be considered a fuel, since in many cases it provides most of the heat required to raise the gases to the thermal oxidizer operating temperature. In this sense, schemes that inject the waste stream downstream of the burner are staged fuel concepts, although they are not generally thought of as such. In contrast to auxiliary fuel staging, a diluent is usually not required with a waste stream. With their typically low heating value, chances of flame formation at the injection point are minimal. Usually the burner is operated at relatively high excess rates to provide the oxygen necessary to oxidize the VOCs injected with the waste stream. This has the added benefit of lowering NOx emissions from the burner itself. An example of fuel staging in this manner is shown in Figure 11.8.

In a third form of fuel staging, the heating value of the waste stream is enhanced by adding auxiliary fuel to it. The waste stream now contributes a higher proportion

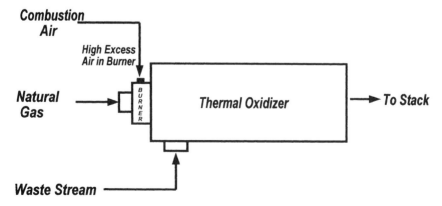

Figure 11.8 Fuel staging with "rich" waste stream.

of heat to the system, and the auxiliary fuel burner firing rate is reduced. The auxiliary fuel added to the waste stream is oxidized rather than combusted. The distinction made here is that oxidation occurs without a flame, whereas combustion produces a high temperature (and high NOx-producing) flame.

Another form of fuel staging is similar in concept to "reburning," a design approach being used in utility boilers to reduce NOx emissions. Here, auxiliary fuel is injected downstream of the primary combustion zone in such quantities as to consume all of the oxygen present and generate an oxygen-deficient (reducing) environment to reduce any NOx present. Secondary air is then added downstream to consume the reducing gases and provide excess oxygen. A schematic of this concept as applied to thermal oxidation is shown in Figure 11.9, where the waste stream acts as a fuel for consuming excess oxygen from the burner.

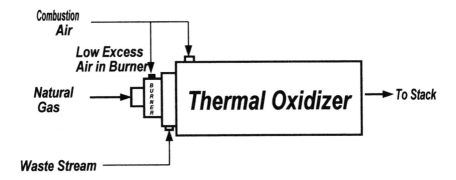

Figure 11.9   "Reburning" concept for NOx reduction.

11.14   STAGED AIR OXIDATION FOR CHEMICALLY BOUND NITROGEN

The aforementioned techniques for reducing thermal NOx generation in a combustion system do not apply to waste streams containing chemically bound (also called fuel bound) nitrogen compounds. Examples of such compounds are dimethyl amine (C_2H_7N), acrylonitrile (C_3H_3N), acetonitrile (C_2H_3N), pyridine (C_5H_5N), and aniline (C_6H_7N). Many industrial off-gases contain these and other chemically bound nitrogen species. When oxidized in a single-stage thermal oxidizer, very high levels of NOx can be produced, depending on their concentration in the waste gas stream. Table 11.1 shows estimates of conversion of this nitrogen to NOx.

To minimize NOx formation from chemically bound nitrogen, an air-staged thermal oxidizer can be applied. The air-staged thermal oxidizer is similar in concept to the air-staged burner except that a residence chamber separates the air injection points. This is shown diagrammatically in Figure 11.10. This concept is used both with and without cooling of the reducing zone gases before second-stage air injection.

The purpose of the interstage cooling is to prevent thermal NOx formation when secondary air is added. The reducing gases can be at a high temperature (>2000°F)

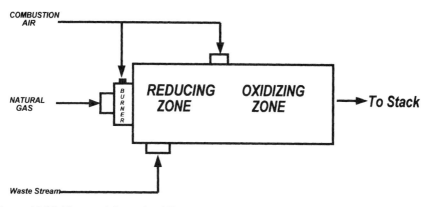

Figure 11.10 Air-staged thermal oxidizer to treat waste streams containing chemically bound nitrogen.

and contain combustible species (CO and H_2). Potentially, a high temperature flame front can form when the secondary air is added, generating thermal NOx. Interstage cooling minimizes this potential. However, if the reducing zone gases are at a lower temperature, interstage cooling could prevent their oxidation once air is added because they (CO and H_2) may be below their autoignition temperature. Interstage cooling is achieved in practice with boiler tubes, heat exchangers, flue gas recirculation, or water injection.

With this staged air concept, the quantity of air injected into the reducing zone is less than the amount theoretically required (stoichiometric). In this oxygen-deficient zone, most of the nitrogen in a nitrogen-containing chemical species is converted to molecular nitrogen gas (N_2) rather than NOx. The exact conversion rate is a function of the concentration of chemically bound nitrogen, the reducing zone stoichiometry, the reducing zone temperature, and the reducing zone residence time.

The optimum conditions required to minimize NOx formation vary with the waste stream. The effect of reducing zone stoichiometry is shown in Figure 11.11. The optimum stoichiometry is typically in the range of 0.5 to 0.8. This value represents the fraction of stoichiometric air required and is called **lambda** (λ) by many in the industry. Others prefer to use the term **equivalence ratio** (φ) which is the inverse of the lambda ($1/\lambda$).

The higher the temperature in the reducing zone, the less NOx produced from the chemically bound nitrogen species present. However, there is a direct relationship between temperature and the reducing zone stoichiometry (λ). The greater the proportion of combustible components consumed in the reducing zone, the higher the temperature. However, this also increases the reducing zone stoichiometry. In practice, the reducing zone stoichiometry is varied by varying the primary air injection rate to optimize NOx emissions. During the air-staged thermal oxidizer design, the temperature of the reducing zone is predicted for various stoichiometries using chemical equilibrium computer codes. The minimum temperature in this zone should be at least 1600°F to ensure some NOx reduction. In fact, temperatures lower than 1500°F are usually not sustainable because they are too close to the autoignition

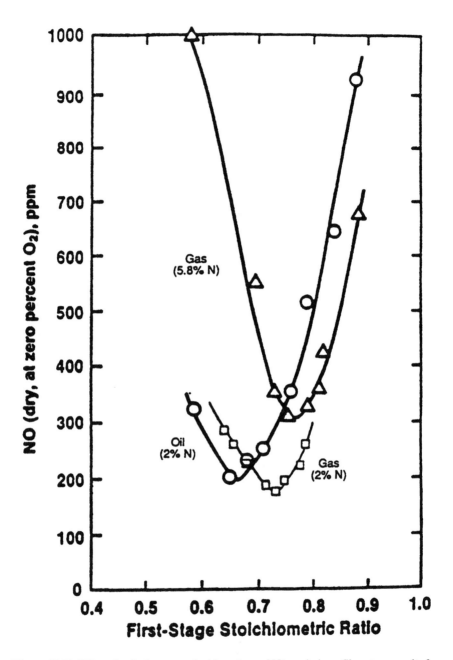

Figure 11.11 Effect of reducing zone stiochiometry on NOx emissions. Shown are results from test firing 2.0 and 5.8% nitrogen gas fuel and 2.0% nitrogen distillate fuel oil/pyridine mixture. Source: "A Low NOx Strategy for Combusting High Nitrogen Content Fuels," EPA-600/7-90-002, January 1990.

temperature of the gases present and the partial combustion required does not occur. The maximum reducing zone temperatures used in these types of systems are only limited by the refractory insulation used. Commercial installations are operating with temperatures approaching 3000°F.

Many industrial off-gases already contain NOx as one of their components. Air-staged thermal oxidation is also effective in reducing NOx that is already present. Reductions of greater than 90% of incoming NOx levels has been achieved in practice in high temperature reducing zones.

NOx emissions in staged oxidation systems are also a function of the gas residence time under reducing conditions. The reducing zone residence time typically varies from 0.5 to 1.0 s. The largest proportion of NOx reduction occurs in the first 0.5 s. The need for the second 0.5 s depends on the final NOx emissions target.

Recuperative heat recovery in the form of combustion air preheat is also applicable to staged thermal oxidation systems. However, unlike the single-stage system, this preheat does not contribute to significantly higher NOx emissions.

The air staging concept is only applicable to waste streams with little or no oxygen in their composition. The principle of this technique rests on partial combustion of the organic components of the waste stream under oxygen-deficient (reducing) conditions. If a high concentration of oxygen is present in the waste stream, it is not possible to generate a reducing environment unless excessive quantities of auxiliary fuel are added to consume the oxygen. Generally, postcombustion NOx controls must be used with these waste gases if NOx emissions exceed environmental limits.

11.15   EFFECT OF SULFUR

Sulfur species in auxiliary fuels or in the waste stream (e.g., mercaptans) can effect the amount of NOx formed. In fuel-lean systems (excess air), these sulfur compounds tend to lower NOx emissions. Under fuel rich (reducing) conditions, the effect is small when the stoichiometry is in the higher end of the range (~0.8). But at lower stoichiometries (e.g., $\lambda = 0.5$), these sulfur compounds tend to increase NOx emissions. Thus, staged thermal oxidation systems for control of chemically bound nitrogen can be affected by the presence of sulfur species.

12 Postcombustion NOx Control

CONTENTS

Combustion control techniques are the most cost-effective for control of NOx emissions from combustion systems. As acceptable levels of NOx emissions are lowered, even these techniques may not be sufficient to meet environmental regulations. In other cases, combustion control techniques may not be appropriate. For contaminated air streams containing nitrogen-bearing compounds, air staging is not applicable because the oxygen in the waste stream makes reducing conditions unattainable. In these circumstances, postcombustion NOx control may be the only alternative.

Two types of postcombustion NOx control are available: selective noncatalytic reduction (SNCR) and selective catalytic reduction (SCR). In SNCR, a reagent, usually ammonia or urea, is injected into the hot thermal oxidizer combustion products. The reagent reacts with the NOx present (predominantly NO form) and converts it to nitrogen gas. No catalyst is required for this process. It is driven by the high temperatures normally found in thermal oxidation systems. SCR is similar to SNCR except that the reagent is injected upstream of a catalyst bed and at a much lower temperature. Both techniques are dependent on intimate mixing of the reagent with the NOx in the combustion products for maximum effectiveness. Both techniques are termed "selective." In this context, "selective" refers to the fact that the reagent injected preferentially reacts with the NOx present in the combustion products, rather than the oxygen present.

NOx can also be scrubbed from the combustion products using gas scrubbing systems similar in concept to those used to remove acid gases from the combustion products (see Chapter 13). They consist of multiple packed towers where the nitric oxide form of NOx is oxidized to nitrogen dioxide. A strong oxidizing agent, such as chlorine dioxide is used at a specific pH. Subsequent absorption towers remove the nitrogen dioxide through reaction with a reducing agent such as sodium hydrosulfide, converting it to sodium thiosulfate and nitrogen gas. The disadvantage of these systems is that they create a liquid blowdown stream which usually requires further treatment before disposal. They are not commonly used with thermal oxidation systems and will not be discussed further.

12.1   SELECTIVE NONCATALYTIC REDUCTION (SNCR)

Several reagent chemicals can be used in this application. The first and still a commonly used reagent is ammonia. Its chemical formula is NH_3. Its use for combustion NOx control was first patented by Exxon and called the Thermal DeNOx process. The original Exxon patent has since expired. Its effectiveness and effective temperature range can be enhanced by the presence of hydrogen gas. Hydrogen can be produced by electrolytically disassociating a side-stream of the primary ammonia stream into hydrogen and nitrogen. This stream is then combined with the primary ammonia stream and injected into the combustion off-gases concurrently. While this enhancement has proven effective, ammonia is most often used alone. It can be injected either as an anhydrous vapor or an aqueous liquid.

Urea is another common chemical used in SNCR applications. Its chemical formula is $CO(NH_2)_2$. At high temperatures, it decomposes to form the same intermediate active species as ammonia. Use of urea in SNCR applications is dominated by the NOxOUT version of the process marketed by FuelTech and its licensees. In contrast to ammonia, it is only injected as an aqueous solution.

Although not as widely known as ammonia and urea, cyanuric acid, $(HNCO)_3$, has also been demonstrated in commercial applications for postcombustion NOx control. However, these demonstrations have been primarily for the reduction of NOx from diesel engine off-gas. Cyanuric acid is most effective in a lower temperature range than that of ammonia or urea.

Other chemicals exist which have been shown to be capable of reducing NOx from a combustion off-gas. These include hydrogen cyanide, ammonium carbonate, ammonium sulfate, and dimethyl carbonate. For various reasons, particularly economics, none are used commercially, except where use is justified by an unusual set of circumstances. Therefore, they will not be discussed further.

12.2   CHEMISTRY

The reaction chemistry for the reduction of NOx with ammonia is as follows:

$$4 \text{ NO} + 4 \text{ NH}_3 + \text{O}_2 \rightarrow 4 \text{ N}_2 + 6 \text{ H}_2\text{O}$$

This equation shows that at the exact stoichiometric ratio, 1 lb-mol (or scf) of ammonia is needed for each lb-mol (or scf) of nitric oxide present. Stated otherwise, 0.57 lb of ammonia is needed for each pound of nitric oxide present.

With urea, the reaction chemistry is

$$(\text{CO}(\text{NH}_2)_2) + 2 \text{ NO} + 1/2 \text{ O}_2 \rightarrow 2 \text{ N}_2 + \text{CO}_2 + 2 \text{ H}_2\text{O}$$

In this case, only 0.5 lb-mol (or scf) of urea is needed for each lb-mol (or scf) of nitric oxide present at the exact stoichiometric ratio. However, because of its higher molecular weight in comparison to ammonia, a greater mass of urea is required (1.0 lb) per pound of nitric oxide present.

The cyanuric acid reaction chemistry is complex but can be approximated by the following equation:

$$(\text{HNCO})_3 + 7/2 \text{ NOx} \rightarrow 13/4 \text{ N}_2 + 2 \text{ CO}_2 + 3/2 \text{ H}_2\text{O} + \text{CO}$$

This equation indicates that 0.29 lb-mol (or scf) of cyanuric acid is needed for each lb-mol (or scf) of nitric oxide present. However, because of its relatively high molecular weight, 1.23 lb of cyanuric acid is needed for each pound of nitric oxide present.

The similarities between the NOx reduction chemistries for these three NOx reduction agents are illustrated in Figure 12.1.

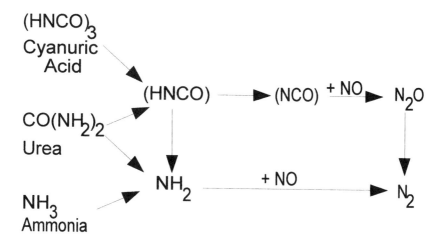

Figure 12.1 NOx reduction chemistry pathways.

12.3 EFFECT OF TEMPERATURE

The effectiveness of each NOx reduction reagent is dependent on the temperature of the combustion products into which they are injected. The temperature window for ammonia is shown in Figure 12.2. At temperatures above 1900°F, the ammonia will actually react with oxygen to form NOx rather than reduce NOx. At low temperatures, the ammonia will not be reactive, and unreacted ammonia, known as slip, will be a constituent of the stack emissions. The optimum temperature for NOx reduction with ammonia is approximately 1750°F.

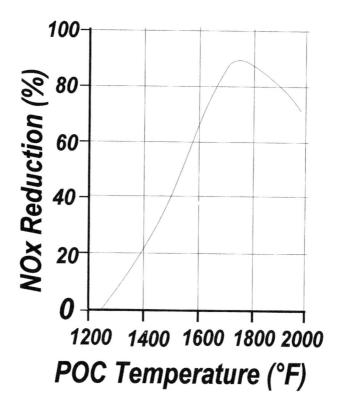

Figure 12.2 Effect of POC temperature on NOx reduction efficiency with ammonia.

A similar temperature relationship for urea is shown in Figure 12.3. It is somewhat broader than the curve for ammonia. Again, too high a temperature will actually result in additional NOx generation rather than NOx reduction. With the NOxOUT process, this temperature window can be broadened even further with proprietary additives. With these additives, urea systems can achieve substantial NOx reduction at temperatures as low as 1400°F.

While ammonia and urea are effective in a similar temperature range, cyanuric acid is most effective at a lower temperature, in the 1350 to 1450°F range. However, substantial NOx reductions can be obtained at temperatures as high as 1700°F.

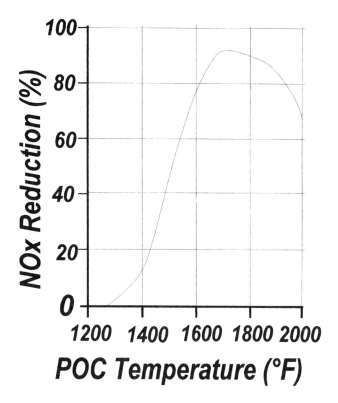

Figure 12.3 Effect of POC temperature on NOx reduction efficiency with urea (no enhancers).

12.4   NORMALIZED STOICHIOMETRIC RATIO

The quantity of reagent injected for NOx reduction is usually stated in terms of the normalized stoichiometric ratio (NSR). A NSR of 1.0 represents the exact amount of reagent needed to satisfy the reaction chemistry. For example, one scf of ammonia added to one scf of nitric oxide is equivalent to a NSR of 1.0. In practice, an NSR greater than 1.0 is required to achieve high NOx reduction efficiencies. Values typically range from 1 to 3. However, there is a limited benefit of increasing the NSR beyond a 2:1 ratio. The stack ammonia slip concentration usually increases as the NSR ratio increases.

12.5   NOx INLET LOADING

The NOx percentage reduction will decrease as the NOx inlet loading decreases. Whereas it may be possible to achieve 70 to 80% NOx reduction when the initial NOx concentration is 200 ppmv or more, it is difficult to achieve the same percentage reduction when the inlet concentration is less than 50 ppmv. Nonetheless, the reduction even from low initial levels may be enough to meet emission limits. For example, a 70% NOx reduction from an initial level of 200 ppmv results in a stack emission of 60 ppmv, while a 40% reduction from 50 ppmv results in a stack emission of 30 ppmv. Because of the relatively small amount of reagent added to the combustion off-gas, the increase in total off-gas volume is insignificant, and percentage reductions can be calculated directly.

12.6 EFFECT OF RESIDENCE TIME

As with all chemical reactions, the NOx reducing reactions require time for reaction. While the reaction chemistry itself is very rapid, intimate mixing of the reagent with the combustion off-gases is crucial to obtaining high NOx reduction efficiencies. In fact, the chemistry of NOx reduction is straightforward. Mixing of reagent with combustion gases represents the most technically challenging aspect of the process. Usually a residence time of 0.25 to 0.5 s is allowed for mixing and reaction kinetics.

12.7 EFFECT OF POC CARBON MONOXIDE
         CONCENTRATION

All combustion off-gases contain some level of CO. With urea, carbon monoxide widens the effective temperature window, shifts the temperature window to a lower regime, lowers the peak removal efficiency, and reduces ammonia slip. With ammonia, CO shifts the effective temperature band to a lower temperature region, has little effect on the width of the temperature band or peak NOx reduction efficiencies, and reduces ammonia slip.

In most well-designed and operated VOC thermal oxidation systems, CO concentrations are less than 100 ppmv. At these levels, the effect of CO on SNCR systems is not significant.

12.8 PRACTICAL REDUCTION LEVELS

Most commercial applications of SNCR NOx reduction technology have been for electric utilities. NOx reductions tend to be on the order of 30 to 40%. These relatively low reduction percentages can be attributed to the fact that these SNCR systems were retrofitted into a boiler system. Since the boiler system was not originally designed for SNCR, it was necessary to "shoehorn" the SNCR system into place. This resulted in a nonoptimum temperature profile, lack of residence time, and nonuniform mixing. The size of utility applications also increases the difficulty of achieving uniform mixing.

Many of these drawbacks can be avoided in thermal oxidation systems. Unlike boilers which have a continuously changing temperature profile as heat is absorbed by the boiler tubes, the temperature in a thermal oxidizer is relatively uniform, at least near the exit. Typically, the size of thermal oxidizers in comparison to utility boilers is much, much smaller. Therefore, the problems with mixing and uniform reagent distribution are greatly diminished. However, mixing is still crucial. Generally, the smaller the unit, the easier to achieve uniform mixing and the greater the NOx reduction efficiency. NOx reduction efficiencies in the range of 70 to 80% can be obtained with ammonia and urea in units with internal diameters under 10 feet. Of course, this is a function of the design and operating parameters previously described. Reduction efficiencies up to 90% have been reported with cyanuric acid.

12.9　INJECTION METHODS

A different injection technique is used for each of the three major reagents used in SNCR applications. A description of the storage, feed, and injection systems for each is described in the following sections.

12.9.1　AMMONIA

Unlike urea and cyanuric acid, ammonia can be injected either as an anhydrous vapor or an aqueous liquid. With anhydrous injection, pure ammonia liquid is stored under pressure. An electric heater is used to vaporize the liquid. As the ammonia vapor is conveyed to the injectors, a carrier gas, typically air or nitrogen, is mixed with the vapor to provide additional mass. The ratio of carrier gas to ammonia vapor can be as high as 20:1. This mass provides the momentum needed for the mixture to penetrate the entire cross section through which combustion off-gases are flowing. The pressure of the combined gases is typically 50 to 60 psig upstream of the injection nozzles. Injection nozzles are located on the circumference or perimeter of the vessel. The number of nozzles required is dependent on the size of the combustion chamber. Sometimes staggered rows of nozzles are provided to enhance mixing. A schematic of a typical anhydrous ammonia injection system is shown in Figure 12.4.

Anhydrous ammonia storage can be a safety issue that many companies would prefer to avoid. An alternative is aqueous ammonia injection. A schematic of this type of system is shown in Figure 12.5. An aqueous solution, typically 25 to 30% by weight, is purchased from a supplier and stored on site. The solution is pumped to the injection nozzles. Flow is controlled with a flow control valve or metering pump. A separate supply of plant water is mixed with the ammonia solution to provide momentum to the liquid spray so that it can penetrate to the center of the combustion gas stream. The flows of aqueous ammonia and dispersing water are controlled independently.

12.9.2　UREA

A schematic of a typical urea injection system is shown in Figure 12.6. It is very similar to the aqueous ammonia injection system. However, a recirculation loop, with a bypass to an electric heater, is included to prevent crystallization of the urea in the tank and maintain a uniform concentration. Specially designed two fluid nozzles are used to provide cross-sectional coverage while maintaining the proper

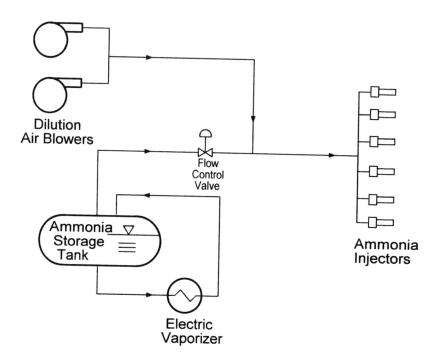

Figure 12.4 Anhydrous ammonia injection system.

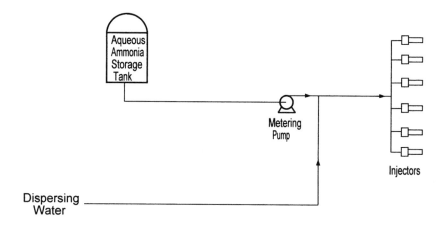

Figure 12.5 Aqueous ammonia injection system.

droplet size. Typically, the urea concentration in the storage tank is approximately 50% by weight. With addition of the dilution water, this concentration is reduced to 2 to 20% as its is injected into the combustion gases.

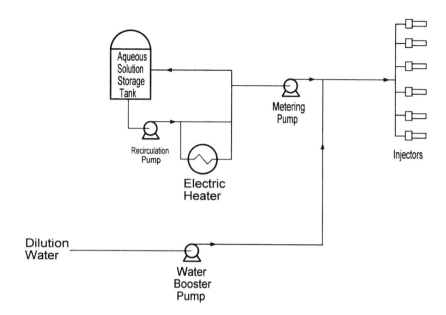

Figure 12.6 Urea injection system.

12.9.3 CYANURIC ACID

Cyanuric acid is a solid at room temperature. It sublimates (solid → vapor) when heated above 700°F. It is not very soluble in cold water. Therefore, the solid form is usually heated until sublimation occurs by using heat from the combustion gases before injecting into the combustion gases. Therefore, it is ultimately injected as a vapor, although its initial form in the storage container is a solid. Typically, the solid is pneumatically conveyed. In one method, part of the conveying line is embedded in the combustion chamber. The sublimation occurs as the reagent is conveyed. Its primary application has been for reducing NOx emissions from large diesel engines.

12.10 COMPUTATIONAL FLUID DYNAMIC MODELING

The chemistry of NOx reduction as described earlier is straightforward. The real technology comes from mixing the reagent with the combustion gas stream. If a reagent molecule does not come in contact with a nitric oxide (NO) molecule, no reaction can take place. The number, location, and design of injection nozzles are crucial to achieving high NOx reduction efficiencies. Testing with small-scale prototypes can provide valuable information. However, while these results can be applied to similarly sized vessels, valid concerns exist when extrapolating this information to vessels significantly larger in cross section or with different flow patterns or velocities.

SNCR system suppliers mitigate these concerns by applying computational fluid dynamics (CFD) to commercial-scale designs. CFD divides a flow area into a large number of cells or control volumes, called a grid or mesh. In each of these cells, Navier-Stokes equations, the partial differential equations that describe fluid flow, are written algebraically, to relate such variables as pressure, velocity, and temperature to values in neighboring cells. These equations are then solved numerically, yielding a picture of the flow corresponding to the level of resolution of the mesh.

CFD techniques can be used to predict temperature, reagent distribution, gas velocities, droplet size, and reagent concentrations as a function of time. This technique is applied to SNCR systems to select the location and number of injection nozzles. Distribution of reagent across the entire cross section of combustion flow is of primary importance. Using this tool, a theoretical representation of this distribution can be generated and the number, location, and size of nozzles can be optimized.

12.11   AMMONIA SLIP

Not every molecule of reagent that is injected reacts with a NOx molecule. A very small fraction usually remains unreacted and appears as part of the stack gas. This unreacted quantity is called "slip." Even if urea is the reagent used, the unreacted portion will appear as ammonia. Ammonia slip is a function of the normalized stoichiometric ratio (NSR), temperature of combustion off-gases, thoroughness of mixing, and type of reagent used. The higher the NSR ratio, the more unreacted reagent is available to become slip. While slip increases with NSR ratio, not all unreacted reagent becomes slip. Some proportion reacts to form NOx, while some of the remainder reduces this NOx formed. Ammonia tends to produce the most slip, cyanuric acid the least, and urea intermediate levels.

Theory suggests that ammonia slip will decrease at higher CO concentrations because the CO oxidation reactions increase local radical concentrations that enhance ammonia reactions. Urea generates a small amount of carbon monoxide by virtue of the fact that not all of the CO-portion of the urea molecule is oxidized when urea is injected into a combustion off-gas. This partially explains its lower slip. Some of the urea also follows the cyanuric acid pathway shown in Figure 12.1, where NOx is reduced without ammonia formation. Injection of methanol along with the NOx reduction reagent also reduces ammonia slip. Most modern SNCR systems can control ammonia slip to less than 10 ppmv.

12.12   REAGENT BY-PRODUCTS

As discussed in previous chapters of this book, VOC compounds containing a halogen or sulfur atom can produce acid gases in a thermal oxidation system. Their potentially corrosive effect on refractory and metals has also been described earlier. They can also be problematic with SNCR systems. When sulfur is present, both ammonium sulfate $(NH_4)_2SO_4$ and ammonium bisulfate (NH_4HSO_4) can form. Both salts are brownish gray to white in color and soluble in water. Ammonium bisulfate is a sticky substance that can form deposits on lower temperature sections of heat exchangers or waste heat boiler components. It causes rapid corrosion of metals as well as fouling and plugging. Ammonium sulfate is not corrosive, but its formation contributes to fouling and plugging and increased particulate emissions.

When hydrogen chloride (HCl) is a product of the thermal oxidation reaction, this HCl will react with ammonia to form ammonium chloride. Ammonium chloride is a white, crystalline solid that is water soluble, hygroscopic, and corrosive. If present in a significant concentration, it will form a detached, white plume in the stack exhaust. An opacity monitor will not detect this plume.

The fact that hydrogen chloride reacts with ammonia has been used as a technique for removing HCl from combustion products. Ammonia is injected into thermal oxidizer products of combustion, the gases quenched, and the resulting solid ammonium chloride particulate collected in a baghouse.

Testing has shown that not all NO is converted to molecular nitrogen gas in a SNCR process. Some nitrous oxide (N_2O) is also formed. The conversion rate is not a function of the quantity of reagent used, but rather the type of reagent used. With ammonia, the conversion to nitrous oxide is less than 5%. However, with urea, N_2O can represent up to 25% of the NOx reduction, and with cyanuric acid the conversion can be as high as 40%. Even though N_2O is not considered a component of NOx, it is a greenhouse gas. While its emissions are not currently regulated, that could change in the future.

12.13   SELECTIVE CATALYTIC REDUCTION (SCR)

Selective catalytic reduction is similar to selective noncatalytic reduction in that the same chemical reagents are used to reduce the concentration of NOx in the products of combustion. However, with SCR, these reagents are injected upstream of a catalyst bed and after the POC has been cooled. The catalysts promotes the same chemical reactions as SNCR, albeit at a lower temperature and a higher conversion rate. SCR is attractive because it can routinely achieve NOx reductions up to 90%. To achieve these NOx reduction levels, the same parameters affecting SNCR optimization also apply to SCR, including temperature regime, residence time (i.e., gas hourly space velocity [GHSV]), and mixing. With SCR, the POC must be oxidizing (contain oxygen). In fact, NOx reduction efficiency will suffer if the oxygen concentration of the POC is less than 2 to 3%. Nonselective catalytic reduction (NSCR) is a similar NOx reduction technology applied to oxygen-deficient gas streams.

Ammonia is the most common reagent used in SCR systems. However, urea can also be used. However, cyanuric acid is not applicable. Because of the health and safety risks posed by anhydrous ammonia (a pressurized liquid), smaller SCR systems use aqueous ammonia injection. Larger systems use anydrous injection, since it is usually the most cost-effective approach when large quantities are required.

The use of catalysts has been discussed previously for oxidation of VOCs. Direct catalytic oxidation of VOCs was discussed in Chapter 9, and Chapter 10 discussed regenerative catalytic oxidation (RCO). The principles and concerns with use of catalysts discussed in those chapters also apply to SCR systems.

Two geometrical configurations of catalyst substrate are used in SCR applications: plates and honeycombs. Plate-type catalysts consist of flat plates separated by spacers. The honeycomb substrate is the same as the structured packing used in regenerative thermal oxidizers (RTOs) and as the catalyst substrate used for direct and regenerative catalytic oxidation (RCO) and shown in Figure 10.2. In SCR applications, the length of the honeycomb is typically 1 m. Grid openings (called pitch) are typically 4 to 8 mm. Smaller cells are used in clean gas applications, while the larger cells are used for "dirty" gas applications. A schematic of a SCR system is shown in Figure 12.7.

The catalyst housing is usually rectangular to accommodate the rectangular catalyst elements. The catalyst is typically arranged in the housing in a series of two to four beds. It is common to include provisions for an additional bed that is not installed initially. This arrangement provides a safety factor to allow for catalyst aging or improved NOx reduction by adding another catalyst bed. The quantity of catalyst required is dependent on the reaction kinetics for that particular type of catalyst and the degree of NOx reduction required.

Several different types of catalysts are available for use at different gas temperatures. Base metal catalysts such as titanium and vanadium oxides are effective when injected into combustion products in the 450 to 800°F temperature range. For higher temperature operation, 675 to 1000°F, zeolite-based catalysts can be used. Precious metal catalysts such as platinum and palladium are available for low temperature operation (350 to 550°F).

With SNCR systems, most of the excess reagent injected tends to oxidize at the high gas temperatures. However, with SCR systems, the gas temperature at which the reagent is injected is usually too low to oxidize the excess. Therefore, any excess reagent will appear in the stack gas as ammonia slip. With SCR systems, reagent is injected at or near the exact stoichiometric ratio, since any excess will generate ammonia slip.

SCR systems are used even when the POC contains acid gases. This is best demonstrated by the fact that SCR systems are in operation on coal-fired utility boilers that burn coal with sulfur contents as high as 3%.

To operate at these high sulfur loadings, ammonia slip must be minimized. Otherwise, the unreacted ammonia will react with sulfur trioxide and form ammonium bisulfate which can mask the SCR catalyst. The sulfur trioxide concentration in the combustion products is increased beyond the amount normally present, because the SCR catalyst converts a small portion of the sulfur dioxide present to sulfur trioxide. By minimizing ammonia slip and suppressing sulfur dioxide oxidation across the catalyst, the amount of ammonium bisulfate formed can be maintained below a level that causes problems.

Hybrids of SNCR and SCR systems are also in use. The SNCR system is installed upstream to remove the bulk of the NOx, while a downstream SCR catalyst performs the final NOx removal. This reduces the size and cost of the SCR system. Ammonia slip generated by the SNCR portion of the process is no longer a concern, since it will act as a reagent for the SCR portion of the process. Both systems can share a common reagent storage and transfer system. For example, if the SNCR system

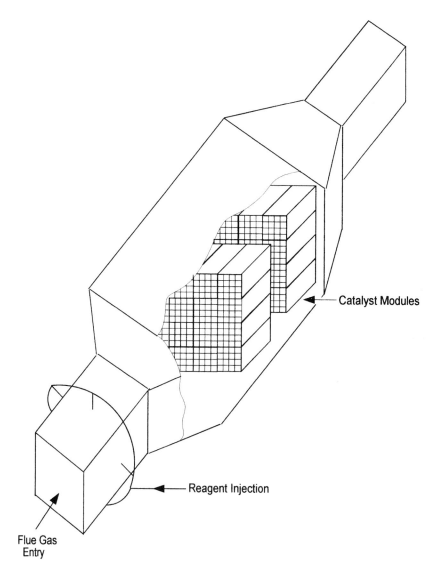

Catalyst Modules

Reagent Injection

Flue Gas
Entry

Figure 12.7 SCR NOx reduction system.

achieves a 70% NOx reduction and the SCR system a 90% NOx reduction, the overall NOx reduction efficiency will be 97%.

The use of SCR systems is not common with VOC thermal oxidation systems. However, as NOx emission limits are lowered, they may become more prevalent. The POC from the thermal oxidizer must be cooled to the optimum temperature range for the catalyst to be effective. With recuperative heat recovery systems, this cooling is a natural consequence of the heat removed from the combustion products.

With waste heat boilers, the SCR catalyst can be installed at the appropriate location between boiler tubes. However, a thermal oxidation system which includes an SCR catalyst must account for the increased pressure drop through the system. Pressure drops through SCR catalysts are typically 3 to 5 in. w.c.

SCR catalysts can also be used with RCO. This application is rare, since these units are inherently low NOx producers. However, with VOCs containing chemically bound nitrogen, high levels of NOx emissions are produced when oxidized even in a RCO. SCR would be appropriate for that application.

13 Gas Scrubbing Systems

CONTENTS

As discussed to some extent in previous chapters, certain VOCs produce objectionable by-products when thermally oxidized. Most common is hydrogen chloride/chlorine gas and sulfur dioxide/trioxide. Sometimes the POC contains particulate that must also be removed. Typically, products of combustion that contain these components must be treated before the POC can be emitted to the atmosphere. The downstream cleanup of products of combustion is termed gas scrubbing.

There are two major types of gas scrubbing systems: wet and dry. Within these two types are many variations. There will be some differences between equipment provided by different suppliers. Also, the nature of the pollutant species can impact the design of the gas scrubbing system design. For example, particulate removal requires much different equipment than acid gas cleanup.

13.1 WET SCRUBBERS

Wet scrubbers receive their name from the fact that liquids are used as a scrubbing medium and the by-products of the scrubbing system are a liquid solution or slurry. Wet scrubbers can serve two simultaneous functions: particulate collection and acid gas control. Particulate collection relies on inertial or electrostatic forces, as with wet electrostatic precipitators. Physical forces such as pressure drop maximize collection efficiency for a given particle size. Gas emissions are absorbed in devices (towers) that attempt to maximize liquid-to-gas contact to achieve maximum absorption of the pollutants into the liquid phase. Usually the design of the absorption towers attempts to minimize pressure drop and eliminate the possibility of material build-up or plugging.

A schematic of a typical wet scrubbing system is shown in Figure 13.1. Products of combustion enter the scrubber from the thermal oxidizer through interconnecting ductwork at the operating temperature of the oxidizer. The first step in both the wet and dry systems is quenching of these POCs. In dry systems, the gas is quenched to a temperature consistent with the material limits of downstream equipment. However, with wet scrubbers, the gas is quenched to the adiabatic saturation temperature. This is the temperature at which the gas is saturated with water vapor. That is, spraying more water into the gas does not change its water vapor concentration. The physical shape of a venturi quench is shown on the diagram of Figure 13.2 and photographs of actual units are shown in Figures 13.3 and 13.4. However, other types of quench designs can be equally effective. The basic criterion is that water be intimately mixed with the POC such that the gas is saturated (or nearly so) before entering the next stage of the scrubbing system.

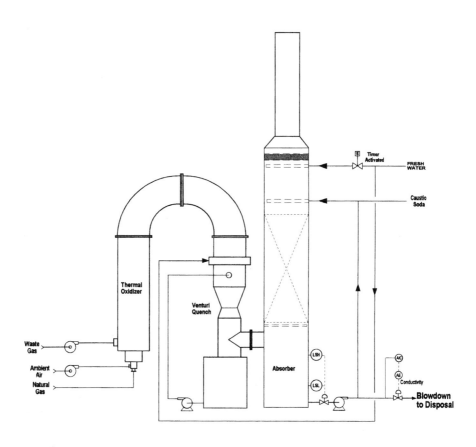

Figure 13.1 Typical wet scrubbing system for acid gas removal.

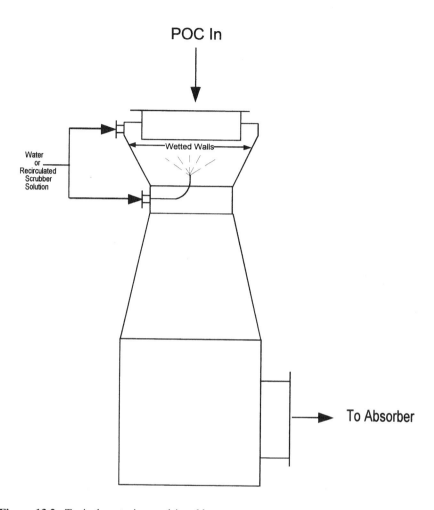

Figure 13.2 Typical venturi quench/scrubber.

After the quenching, the saturated gas enters one or more absorption towers for removal of any acid gases present. The base of the absorption tower(s) may contain a sump providing a level of liquid that is pumped to the top of the tower for recirculation. This recirculation provides a high liquid-to-gas ratio that enhances reaction between chemical reagents in the liquid and acid gases in the thermal oxidizer combustion products. A small portion of the recirculated liquid is removed as a means of discharging the pollutants from the system to maintain the suspended or dissolved solid content of the recirculating solution. This stream is called the "blowdown." Some liquid from the sump is also used in the quench, along with fresh water, to saturate the gas. Fresh chemical reagent is injected into the top of the tower(s) to react with acid gases. Sometimes an induced draft fan is included in the system to provide the motive force to overcome the pressure drop through the system. The fan can be located after the

Figure 13.3 Commercial venturi scrubber installation. (Courtesy of AirPol Inc.)

absorption tower or in-between multiple absorption towers. A photograph of a commercial wet scrubber installation is shown in Figure 13.5.

If a quench tower is used in place of a venturi quench, sufficient time and adequate turbulence must be allowed for the gas to reach saturation. This time is a function of the inlet gas temperature, type of liquid spray nozzle used, and number of nozzles used. Hydraulic nozzles require a greater contact time because they produce coarse droplets that take longer to evaporate. Contact times vary from 1.5

Figure 13.4 Venturi scrubber followed by cyclonic separator. (Courtesy of AilPol Inc.)

to 2 s with high inlet gas temperatures (1500 to 2000°F) to 0.5 s with low inlet gas temperatures (500°F). This low temperature may be a result of an intermediate heat recovery device, such as a waste heat boiler.

Figure 13.5 Gas scrubbing system downstram of RTO. (Courtesy of AirPol Inc.)

13.1.1 ADIABATIC SATURATION

When hot products of combustion enter the gas scrubbing system, the first step in the process is adiabatic saturation (or humidification) of the gas in the quench section. "Adiabatic" refers to the fact that the resulting gas contains the same quantity of heat as the original gas, albeit at a lower temperature.

The ultimate gas temperature reached after saturation depends primarily on the temperature of the gas as it enters the quench section and the initial water vapor concentration of the POC. However, saturation almost always results in a final gas temperature in the range of 140 to 190°F. Once the gas is saturated, further cooling requires a significant amount of energy, since water must be condensed to further reduce the temperature. The saturation temperature can be calculated from the POC composition and initial temperature. Unfortunately, this calculation is not direct. An iterative procedure must be used. This procedure is as follows:[13]

1. Assume a saturation temperature (Ts)
2. Calculate the latent heat of vaporization (Hv) of the water at the assumed adiabatic saturation temperature.

$$Hv = 91.86 \times (705.56 - Ts)^{0.38}$$

3. Calculate the vapor pressure of water (Pv) at the saturation temperature.

$$\log Pv = 15.092 - 5079.6/Ts - 1.6908 \log Ts - 3.193 \times 10^{-3} \times Ts + 1.234 \times 10^{-6} \times Ts^2$$

where

      Ts   = assumed saturation temperature ((R)
      Pv   = water vapor pressure (psia)

4. Calculate the *mean* heat capacity (Cpdg) of the POC (Btu/lb-°F) on a dry (water-free) basis between its initial temperature and the assumed saturation temperature, using methods described in Chapter 5. Calculate the *mean* heat capacity (Cpw) of water vapor (Btu/lb-°F) between the initial POC temperature and the assumed saturation temperature.
5. Calculate the saturation humidity (Sh1)

$$Sh1 = Ht + (Cpdg + Cpw \times Ht) \times (T_1 - Ts)/Hv$$

where

      Sh1   = saturation humidity (lb water vapor/lb dry gas)
      Ht   = actual humidity at temperature T (lb water vapor/lb dry gas)
      T_1 = initial POC temperature (°F)

6. Calculate the saturation humidity (Sh2)

$$Sh2 = 18.016/(M \times (P - Pv))$$

where

      Sh2   = saturation humidity (lb water vapor/lb dry gas)
      M   = POC molecular weight
      P   = system pressure (psia) [usually 14.7]

7. Compare Sh1 to Sh2. Repeat steps 1 through 6 until Sh1 = Sh2.

Example 13.1: The products of combustion from thermally oxidizing a chlorinated VOC at 1750°F are as follows:

	Vol% (Wet)	Vol% (Dry)
Carbon dioxide	3.91	4.91
Water vapor	20.44	0
Nitrogen	64.64	81.25
Oxygen	9.8	12.32
Hydrogen chloride	1.21	1.52

What is the adiabatic saturation temperature after the quench section of a gas scrubbing system?

Using the steps outlined above with an initially assumed saturation temperature of 200°F:

Ts	Ht	T$_1$	Hv	Pv	Cpdg	Cpw	Sh1	Sh2
200	0.157	1750	978.50	11.442	0.269	0.541	0.718	2.154
195	0.157	1750	982.16	10.308	0.269	0.541	0.718	1.439
190	0.157	1750	985.81	9.270	0.269	0.541	0.717	1.047
185	0.157	1750	989.43	8.320	0.268	0.540	0.717	0.799
184	0.157	1750	990.15	8.140	0.268	0.540	0.717	0.761
183	0.157	1750	990.87	7.964	0.268	0.540	0.716	0.725
182	0.157	1750	991.59	7.791	0.268	0.540	0.716	0.691

The adiabatic saturation temperature is between 182 and 183°F.

13.1.2 PARTICULATE REMOVAL

Particulate can emanate from two sources in a thermal oxidation system: inorganic particulate already present in the VOC waste stream or particulate generated during the thermal oxidation reactions. Generally, organic particulate in the waste stream will be combusted to a large extent if the thermal oxidizer is operated at a high temperature (e.g., 1800°F). Of course, the size of inorganic particulate in the waste stream is subject to the conditions under which it was generated. Usually particulate generated from the combustion reactions is submicron in size and requires a high-energy particulate collection system for its removal. Particulate that becomes molten when it is formed in the combustion reactions is the most troublesome. In this case, the POC is typically quenched to below the melting point of the particulate before entering the scrubbing system.

Particulate collectors rely on inertial forces to encapsulate the particulate present in liquid droplets. Impaction is the most common technique applied. The contaminant is accelerated and impacted onto a surface or into a liquid droplet. The kinetic energy of the particle momentum is used to penetrate the surface tension of the scrubbing liquid. Some devices accelerate the particle directly into a liquid film, while others produce a spray of water that allows collection of the particle on the droplet surface.

Particles that escape impact may be captured by interception. Here, the particles do not directly impact the droplets but glance at a tangent to the droplet. They still have sufficient energy to be adsorbed by the droplet. High energy venturis and other devices rely on high density sprays of fine droplets to increase particle interception. Interception is most common with submicron particles that tend to follow gas streamlines.

The smaller the size of the particulate, the more difficult it is to remove from the gas stream. Removal of small particles is enhanced through smaller, more dense, liquid droplets. These smaller droplets are produced by imparting energy to the droplets. This can be done in several ways. One is to reduce the diameter of the throat of a venturi scrubber and thus increase its pressure drop. A second method is to use high pressure drop nozzles to inject the liquid spray. With all methods, the effectiveness of the particle removal is proportional to the power (energy) supplied to the removal device. Thus, high removal efficiencies or removal of very small particles require a substantial energy input. Venturi scrubbers with pressure drops as high as 50 in. w.c. or more are in operation. This pressure drop translates to a high horsepower input for fans in the system.

After the saturated gas leaves the particulate wetting device, it may proceed to a cyclonic entrainment separator where most of the liquid is removed from the gas stream by centrifugal force. This is sometimes built into the lower section of a downstream absorption tower. The gas enters at a tangent to the diameter. Centrifugal forces spin the relatively heavy droplets to the wall, where they separate from the gas stream and drain to the liquid sump below.

13.1.3 Acid Gas Removal Chemistry

If the POC contains acid gas instead of or in addition to particulate, it is usually directed to absorption towers for its removal before emitting the remainder of the POC to the atmosphere. There, reagent solutions are injected which neutralize these acid gases. These reagents are almost always a sodium or calcium alkali compound. The most common are sodium hydroxide, sodium carbonate, and calcium hydroxide (hydrated lime). The reaction chemistry is shown below for various acid gas components:

Sodium Hydroxide (NaOH)

$$HCl + NaOH \rightarrow NaCl + H_2O$$
$$Cl_2 + 2\ NaOH \rightarrow NaCl + NaOCl + H_2O$$
$$SO_2 + 2\ NaOH \rightarrow Na_2SO_3 + H_2O$$

Sodium Carbonate (Na_2CO_3)

$$Na_2CO_3 + H_2O + 2\ SO_2 \rightarrow 2\ NaHSO_3 + CO_2$$
$$Na_2CO_3 + 2\ HCl \rightarrow 2\ NaCl + CO_2 + H_2O$$

Calcium Hydroxide (Ca $(OH)_2$)

$$Ca(OH)_2 + SO_2 \rightarrow CaSO_3 + H_2O$$
$$Ca(OH)_2 + 2\ HCl \rightarrow CaCl_2 + 2\ H_2O$$
$$2\ Ca(OH)_2 + 2\ Cl_2 \rightarrow CaCl_2 + Ca(OCl)_2 + 2\ H_2O$$

Other halogenated acid gases react similarly to hydrogen chloride. Sodium hydroxide is usually purchased as a 50% solution (by weight). Many facilities dilute this concentration to 20%, since this concentration has the lowest freezing point. Sodium carbonate (also called soda ash) is purchased as a solid and must be dissolved in water. Lime is purchased as a solid and must be "slaked," forming a slurry before injection.

13.1.4   Gas Absorption Towers

Gas absorption towers can be divided into two categories: those that disperse liquid over a plate or packing, and those that create a spray of droplets. Both systems rely on the creation of a large liquid surface area. In all cases, the gas enters the bottom of the tower and proceeds upward. Intimate mixing occurs between the liquid and gas over the packing or plates or with the fine spray of liquid in the open spray tower.

Packed bed absorption towers are usually filled with randomly oriented packing material, such as saddles or specially designed shapes that produce high liquid-to-gas surface area. Most applications use plastic materials such as polypropylene. Both fresh reagent and recirculated liquid from the tower sump are distributed over the top of the packing. As the liquid flows through the bed, it wets the packing material and provides interfacial surface area for mass transfer with the gas phase. The rate of mass transfer is directly proportional to the concentration gradient between the gas and liquid phases. The primary design variables for gas absorption are depth of packing, liquid-to-gas ratio, superficial gas velocity, and contact time.

Plate towers contain plates or trays spaced at intervals along the length of the tower. The fresh reagent and recirculated liquid are injected on the top plate. The mixture flows successively across each plate as it moves downward. The saturated gas stream enters the bottom of the tower and flows through openings in each plate. Gas absorption is promoted through formation of gas bubbles that pass through the liquid on each plate. The primary design variables for plate towers are number of plates, liquid-to-gas ratio, and contact time.

In a system used to remove hydrogen chloride from a gas stream, it must be remembered that the HCl releases heat at a rate of approximately 800 Btu/lb as it is absorbed in water. It releases further heat when neutralized by an alkali agent, almost 3300 Btu/lb, for example, when neutralized with sodium hydroxide. In systems with high concentrations of HCl, this heat can be very significant and can require cooling of recycled solutions.

13.1.5   HCl Recovery

Hydrogen chloride is very soluble in water and is easily removed in a wet gas scrubbing system. In fact, HCl removal efficiencies as high as 99% can be achieved in an absorption tower with water only. No chemical reagent is required. This fact is sometimes used to recover the relatively pure HCl solution produced for use in other parts of an operating plant. The concentration generated is controlled by the quantity of fresh make-up water added along with the rate of blowdown. Of course, a small amount of chlorine gas is produced along with the HCl, according to chemical equilibrium described in Chapter 5. Cl_2 cannot be removed with water alone. If the release of this Cl_2 is objectionable, a second scrubbing tower can be added after the first. Water is used in the first stage for HCl removal and recovery, while Cl_2 is scrubbed in the second tower with alkali reagent injection. This configuration is shown in Figure 13.6.

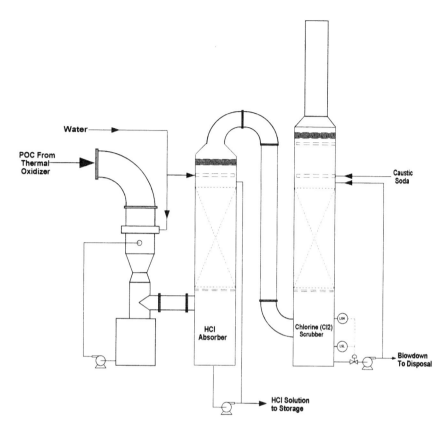

Figure 13.6 HCl recovery scrubber system.

13.1.6 MIST ELIMINATORS

Before the clean gas is exhausted to the stack and ultimately to the atmosphere, it proceeds through a mist eliminator installed near the gas outlet of the absorption tower. Mist eliminators are also called "entrainment separators" and "demisters." Their purpose is to remove liquid droplets entrained in the gas stream. Failure to include a mist eliminator in the scrubber design has resulted in scrubbers emitting more particles than were in the original gas stream. Mist eliminators can remove 99 to 99.9% of the entrained droplets. There are numerous configurations of mist eliminators. Most operate by inertial impaction or centrifugal force. The configuration of one common design, called the chevron type, is shown in Figure 13.7.

13.1.7 MATERIALS OF CONSTRUCTION

Materials of construction of the gas scrubbing system are generally different than the thermal oxidizer. One of the most severe environments encountered is that of the quench chamber. It operates with gas temperatures up to 2200°F and with highly

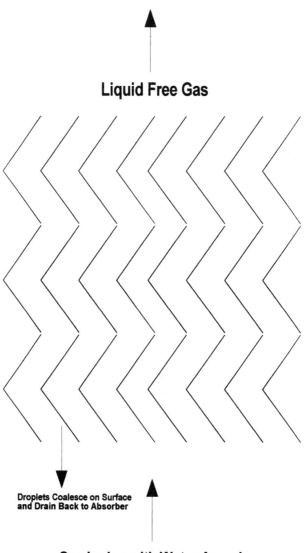

Figure 13.7 Chevron mist eliminator.

corrosive gases. Therefore, the quench is usually designed with a high nickel alloy (e.g., C276, 6XN). The quench must also be designed for the thermal expansion effects of the high temperature differential between inlet and outlet and movement during start-up. Most problems in this area occur at the wet–dry interface, the transition between the liquid and gas phases. In addition to corrosion concerns, this area is frequently where solids buildup can occur.

Once the gas temperature has been lowered through saturation, less expensive materials of construction can be used. Absorption towers constructed of fiberglass reinforced plastic are common. However, provisions must be included in the design to prevent a catastrophic "meltdown" of the tower should a loss of liquid occur through a pump failure or loss of electrical power. A backup supply of fresh water from plant emergency water or a head tank is usually included in the design. It is also common to use plastic or fiberglass piping for circulating liquids because of its low cost and chemical resistance.

13.1.8   Ionizing Wet Scrubber

For removal of very fine (submicron) particulate, such as that produced in thermal oxidation of waste streams containing inorganic or organometallic species, another type of wet scrubber may be used. The ionizing wet scrubber (IWS) combines the principles of electrostatic particle discharging, inertial impaction, and gas absorption. With IWS, high-voltage ionization is used to electrostatically charge particles before they enter a packed-bed scrubber section. In the packed bed, particulate is removed by attraction of charged particles to neutral surfaces. The packing also acts as an impingement surface. The electrostatically charged particles are attracted to the neutral surfaces by a phenomenon called "image force attraction." A recycled water or reagent is injected over the top of the packing. The particles are carried along with this liquid and are removed with the liquid discharge. Gas flow through an IWS system is typically horizontal. IWS systems can remove particulate as small as 0.1 micron.

13.1.9   Wet ESP

Electrostatic precipitators (ESPs) use an electrostatic field to place a charge on particulate matter. These charged particles are then collected on a surface of opposite charge. ESPs are very effective in removing submicron particles. ESPs can be either dry or wet. The dry ESP will be discussed in the next section. The wet ESP can be configured with horizontal or vertical gas flow. Their shape can vary from square to concentric. Concentric ESPs use high intensity ionization electrodes mounted on a central tube to produce a high intensity corona. The electrodes are energized by a transformer/rectifier set. As the gas enters the tube, it passes through a charging zone located at the top section of the tube. The charged particles are attracted to the tube walls by the electrostatic field created by the central support tube as the gas flows down the tube. The cleaned gases exit through a bottom outlet plenum.

Square ESPs use the same attraction principle, but use an alternating array of negatively charged wires or grids and positively charged, flat collection plates. The discharge electrodes can be metal wires tensioned by weights or rigid frames.

The wet ESP is termed such because it uses a flow of liquid to wash the particulate from the collecting electrode. In one approach, water sprays are used. In another, the POC is saturated (and sometimes subcooled) upstream of the wet ESP. Water vapor then condenses on the particulate collection tubes and carries away the particulate as it drains to the bottom. A backup water flushing system can also be included in the design. While water is used as a flushing medium, it also serves to ensure that the particle is capable of receiving and retaining an electric charge regardless of the dielectric nature of the particle. An electric charge is placed on the water droplet itself. These droplets are large compared to the particulate. The water droplets attract the particulate and are themselves attracted to the collector plate. Some designs require a mist eliminator at the outlet.

13.2   DRY SYSTEMS

There are a variety of dry scrubbing systems that can be used downstream of a thermal oxidation system to collect both particulates and acid gases. The most common uses a spray dryer, followed by a fabric filter (baghouse). Sometimes a dry ESP is used in place of the baghouse. This configuration is illustrated in Figure 13.8.

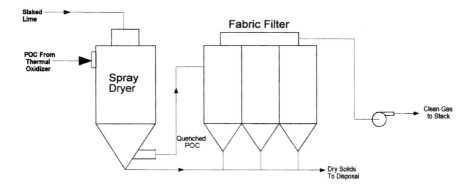

Figure 13.8 Typical dry scrubbing system.

The first device in this type of configuration is the spray dryer. Here the gas is quenched from its temperature exiting the thermal oxidizer with a liquid spray into the spray dryer. Usually, this liquid contains the reagent needed to react with and remove any acid gases present. More often than not, hydrated (slaked) lime slurry is used. It is injected as a 5 to 50% slurry. A water dilution line controls the slurry strength.

The slurry is then atomized into a fine spray as it is injected, using high-pressure hydraulic, two-fluid air atomized nozzles, or a rotary atomizer. The spray dryer is sized to produce a residence time of 5 to 10 s. The POC usually enters around the atomized liquid and prevents it from contacting the dryer walls. The water evaporates from the slurry, cooling the POC and exposing the lime reagent to the acid gases for reaction. Unlike the wet scrubber, the gases are only cooled to a temperature dictated by the downstream particulate collection device. A typical temperature is 350°F. The lime reaction, as described previously, produces a particulate itself that is then removed, for example, in a baghouse.

Operating conditions influence spray dryer acid gas removal efficiency. Performance is particularly affected by approach temperature. The approach temperature is the difference between the operating temperature and the adiabatic saturation temperature. A stoichiometric ratio of lime (calcium) to acid gas greater than one is required for high removal efficiencies. Typical values are in the range of 1.25 to 1.50. Increasing the ratio beyond 2.0 has a diminishing effect on performance. Approach temperatures can be as low as 20°F. However, low approach temperatures can present a corrosion problem if chlorine is present. Calcium chloride is a deliquescent solid. This means that it absorbs moisture and then dissolves in the absorbed moisture. Solids that contain calcium chloride become sticky and can lead to accumulation and plugging of the discharge. To avoid this problem, the approach temperature is normally set to 50 to 100°F above adiabatic saturation when chlorine is known to be present. On the other hand, the presence of chlorine has been shown to increase SO_2 removal efficiency when both are present in combustion products.

Dry scrubbers are not as effective for acid gas removal as wet scrubbers. Whereas 99.99% hydrogen chloride removal is possible in a wet scrubber, dry scrubber HCl removal efficiencies are generally on the order of 90 to 95%. Wet scrubbers can remove approximately 95% of the sulfur dioxide in POC vs. approximately 75 to 85% for a dry scrubber.

13.2.1 Fabric Filter (Baghouse)

In a baghouse, solids entrained in the spray dryer off-gas form a cake or layer of solids on the surface of the bags. In fact, some additional removal of acid gases from the gas stream occurs as the gas flows through this cake. As this layer increases in thickness, a pulse of gas from the opposite side causes the bags to flex and discharge the solids to a hopper at the bottom of the baghouse. This "blowback" is either activated by a timer or when the pressure drop across the bags reaches a prescribed limit. In "reverse air" systems, the gas used for blowback is POC taken from the outlet of the fan, downstream of the fabric filter. With "pulse-jet" cleaning, a short, high-pressure burst of air is discharged into the downstream side of the bags. Many times this air is blown through a venturi nozzle at the top of each bag to inspirate additional gas from the exhaust plenum. Some systems use mechanical rappers on the bags to dislodge the particulate.

Typical pressure drops across bag filters are 4 to 10 in. w.c. Solids are removed through a rotary valve mechanism at the bottom of the hopper section of the baghouse. The advantage of this system compared to a wet scrubber is that the by-products are in a solid form of relatively low volume. In many cases, these solids do not require further treatment and can be disposed in a landfill.

The term fabric filter and baghouse are used interchangeably. Collection of particles occurs through filtration as a gas passes through a fabric. The basic components of the baghouse are the filter medium (bags), a cage to support the bags, a housing, and a mechanism to discharge accumulated solids. A hopper is included at the bottom.

The gas enters the baghouse near the bottom and then flows to one of several compartments. These compartments contain a bank of bags. The bags hang from a tubesheet that separates the clean and dirty gases. By using more than one compartment of bags, the bags can be cleaned without interrupting gas flow through the system. For example, in a four-compartment baghouse, one compartment may be momentarily off-line during blowback, but flow continues through the other three compartments.

One of the most important design parameters of the fabric filter is the air-to-cloth ratio. This is the volumetric flowrate of the flue gas (ft³/min) divided by the surface area of the fabric (ft²). The higher this ratio, the smaller the fabric filter, but the higher the pressure drop. Reverse-air (flue gas) and shaker baghouses with woven fabric bags generally have air-to-cloth ratios ranging from 2.0 to 3.5. Baghouses utilizing the pulse-jet (compressed air) cleaning technique usually use felted fabric bags and have air-to-cloth ratios from 5 to 12.

13.2.2 DRY ESP

Wet ESPs were discussed previously. Dry EPSs operate using the same basic concept. A high voltage is applied between an array of negatively charged wires or grids and positively charged collection plates. This produces an electrostatic field. In the space between the electrodes, a corona is established around the negatively charged electrode. As the particle-laden gas passes through this space, a corona ionizes molecules of the electronegative gases present in the gas stream. These molecules attach themselves to particulate matter entrained in the gas stream, charging the particles. The charged particles migrate to the oppositely polarized collection plates. The electrical field that drives the charged particles is created by applying a high DC voltage (typically 15,000 to 50,000 V) between the corona-generating electrode and either flat or concentric cylindrical collecting plates. The strong electrostatic field inhibits re-entrainment. Mechanical rappers are used to vibrate the plates and dislodge the particles. They fall into a hopper below from which they are removed. Since no liquids are used in the process, the particulate is removed as a dry solid; thus, the term "dry ESP."

Most ESPs are of the plate-wire design. In older ESP units, the high voltage corona-generating electrodes are long wires (either plain or barbed) which are weighted and hang between a bank of flat, parallel collecting plates. In more recent designs, the discharge electrodes are rigid frames, often fitted with barbs or other points. The particulate buildsup as a cake on the collection plates in both designs. Usually the POC flow to the ESP is slowed and straightened to evenly distribute the gas over the bank of collecting plates. The force tending to divert the charged particles to the collection plates is relatively weak and flow disturbances or eddies will inhibit collection efficiency and also re-entrain material that has already been collected. Gas bypassing around the collection zone can also reduce collection efficiency. This includes flow through hoppers and over the top of collection plates. This type of bypassing is called "sneakage" and can be minimized by proper baffling.

Particulate removal efficiency of the dry ESP depends on gas flowrate, temperature, moisture, the electrical resistivity of the particles, the inlet loading, and the particle size distribution. The dry ESP is very effective for particles with resistivities of 10^4 to 10^{10} ohm-cm. Most common industrial particulates exhibit resistivities in this range. Particles with lower resistivities tend to lose their charge when they contact the collection plate and are re-entrained in the flue gas. Particles with higher resistivities are difficult to remove. Removal of particulate from the collection surface is critical to efficient ESP operation. Particulate can act as an insulator, preventing the electrostatic action from occurring and reducing its effectiveness.

The chloride content of the combustion products can also affect ESP performance. The hydrophilic nature of the chloride leads to a higher moisture content. This decreases particle resistivity and increases particle cohesiveness. Lower particle resistivity results in better removal, while higher cohesiveness results in less entrainment. Both translate to better ESP performance.

One of the most important design parameters of the dry ESP is the specific collection area (SCA). This is the area of the collecting electrodes or plates divided by the volumetric flowrate of the gas. For a given application, the higher the specific collection area, the greater the collection efficiency.

Dry ESPs can operate with flue gas temperatures as high as 750°F. Particulate removal efficiencies generally range from 95 to 99%. In fact, dry ESPs are very effective in removing submicron particulate. Advantages of ESPs are their reliability and low maintenance, relatively low power requirements, high collection efficiency over a wide range of particle sizes, and ability to treat a humid gas stream. Disadvantages include a sensitivity to changes in gas properties and particle size distribution, and inability to collect particles with low resistivity. Also, if installed downstream of a spray dryer for acid gas removal, they do not enhance removal efficiency as does the filter cake on the fabric filter.

Characteristics of particulate removal equipment are compared in Table 13.1.

TABLE 13.1
Characteristics of Particulate Removal Equipment

Venturi Scrubber

Greater than 99% removal of particulate in 0.5 to 5 micron range

Particulate collection efficiency proportional to pressure drop (high energy cost for high efficiency)

Ability to treat high temperature gas streams

Unaffected by gas composition

Relatively small footprint required

Requires corrosion resistant alloys (with acid gases)

Wet discharge

Relatively low initial cost

Dry ESP

High collection efficiency for submicron particles

Acceptance of high gas inlet temperatures (~750°F)

Ability to handle large gas flow rates

Low maintenance and high reliability at steady state gas conditions

Low pressure drop

Low power requirement

Dry discharge

High initial cost

Limited gas turndown

Sensitive to flue gas temperature and humidity

continued

TABLE 13.1 (CONTINUED)
Characteristics of Particulate Removal Equipment

Dry ESP (continued

Special precautions for high voltage
Relative ineffectiveness for particles with low resistivities

Ionizing Wet Scrubber

Low energy costs
Medium collection rate of submicron particles, aerosols, and mists
Unaffected by particle size
Relatively high capital cost
Wet discharge

Wet ESP

High collection efficiency for submicron particles
Ability to handle large gas flow rates
Low maintenance
Low pressure drop
Low power requirement
Wet discharge
Smaller than equivalent dry ESP
Unaffected by gas humidity
High initial cost

Fabric Filter

Dry collection
Collection efficiency unaffected by inlet particle loading
Collection efficiency unaffected by gas humidity (but high humidity could "blind" bag)
Gas temperature limited by bag materials
Large size
High pressure drop
High maintenance
Medium capital cost

13.3   HYBRID SYSTEMS

A schematic of a hybrid scrubber system is shown in Figure 13.9 and a photograph of a commerical installation in Figure 13.10. It combines the high acid gas scrubbing efficiency of the wet scrubber with the dry blowdown of the dry scrubber. In this system, scrubber blowdown from the wet scrubber is used to quench the POC in the spray dryer. The solids in the wet scrubber solution (e.g., sodium sulfite), precipitate as the water in which they are dissolved is evaporated. These solids are then collected in the baghouse. The wet gas absorber downstream of the baghouse removes the acid gas from the POC by wet alkali injection. Again, the resulting solution is pumped back to the spray dryer for disposal. Thus, the best features of the wet and dry scrubbing systems for acid gas control are combined.

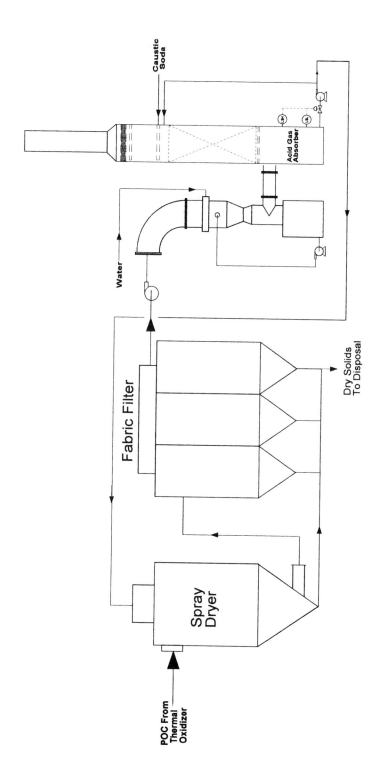

Figure 13.9 Hybrid wet/dry scrubbing system.

Figure 13.10 Photograph of commercial hybrid gas scrubbing system. (Courtesy of AirPol Inc.)

14 Safety Systems

CONTENTS

Thermal oxidation systems can release a significant quantity of energy during the oxidation process. If the system is not designed properly, this energy release could result in harm to operating personnel or damage to equipment. In these systems, safety issues relate primarily to the potential of fire or explosion from auxiliary fuels or from waste streams with high VOC loadings. Common causes of accidents include condensation and collection of VOCs in ductwork, process upsets resulting in unexpectedly high concentrations of VOCs, inadequate or missing safety-related detectors or safety interlocks, improper placement of detectors, and equipment or instrumentation failures.

14.1 LOWER EXPLOSIVE LIMIT (LEL)

The lower explosive limit (LEL) of an organic compound is its minimum concentration in air that will sustain combustion. Conversely, the maximum concentration of a VOC that will sustain combustion in air is called the upper explosive limit (UEL). The LELs and UELs of common organic compounds are listed in Appendix E. LEL data are usually more applicable to thermal oxidation systems since extremely high concentrations of VOC in air are unusual. However, there have been applications where a fuel such as natural gas has been added to a waste stream to increase the mixture concentration above the UEL to prevent a flashback. However, such methods are themselves risky.

The most common accidents with VOC thermal oxidizers involve flashbacks. Here, the VOC mixture in the waste stream rises to a level between its lower and upper explosive limits. A spark, or even the flame in the thermal oxidizer itself, can cause the VOCs to ignite in the ductwork. The flame travels back to the source with such speed and force as to cause an explosion.

Many times a waste stream contains more than one VOC. In that case, the LEL is calculated as follows:

$$\% \text{ LEL of mixture} = \frac{100}{(X_1/L_1 + X_2/L_2 + ...X_n/L_n)}$$

where

$X_1, X_2, ... X_n$ = volume fraction of VOC 1,2, ...n, respectively
$L_1, L_2, ... L_n$ = LEL of VOC 1, 2, ... n, respectively

The LELs of VOCs are typically given in air at 77°F (25°C). However, the LEL decreases with temperature according to the following formula:[14]

$$\text{LEL (at temp t (°C))} = \text{LEL (at 25°C)} \times (1 - 0.000784 \times (t - 25))$$

where

t = actual temperature (°C)

Example 14.1: The LEL of acetone in air is 2.55 vol% at 77°F (25°C). What is its LEL at 400°F (204°C)?

Solution

$$\text{LEL (at 204°C)} = 2.55 \times (1 - 0.000784 \times (204 - 25)) = 2.19 \text{ vol}\%$$

This example indicates that as the temperature rises, a waste stream will enter the explosive range at a lower VOC concentration. While almost all VOC thermal oxidation systems operate at or near ambient pressure, increasing pressure also decreases the LEL. Increasing temperature and pressure also increase the UEL.

A tabulation of the LEL of a particular VOC may not always be readily available. As an approximation, the LEL occurs at 50% of the stoichiometric oxygen concentration at ambient temperature and pressure. The UEL occurs at approximately 3.5 times the stoichiometric oxygen concentration.

Example 14.2: Estimate the LEL of acetone using the 50% stoichiometric oxygen approximation. The chemical formula for acetone is C_3H_6O.

Solution

$$C_3H_6O + 4\,O_2 \rightarrow 3\,CO_2 + 3\,H_2O$$

For each lb-mol of acetone oxidized, $4/0.21 = 19.05$ lb-mol of *air* is required. The stoichiometric concentration of acetone in air is then $1/(19.05 + 1) = 0.0499$ (mole fraction). Half of this value (50% is stoichiometric) is $0.5 \times 0.0499 = 0.025$ mole fraction or 2.5%. This compares very well to the published value of 2.55%.

14.2 MINIMUM OXYGEN CONCENTRATION

As previously discussed in Chapter 6, the ignitibality of a VOC-containing waste stream is inhibited at lower oxygen concentrations. As shown in Table 6.4, with few exceptions, ignition cannot occur if the oxygen content of the waste stream is less than 10%. The important exceptions are waste streams containing hydrogen, carbon monoxide, or xylene. The values in Table 6.4 are recommended values and contain a safety margin (typically 2%) between the recommended value and the actual value at which ignition can occur. If the maximum safe oxygen concentration is not available in the literature, it can be estimated by multiplying the percent LEL by the number of pound-moles of oxygen required for complete oxidation of a pound-mole of VOC. The following example illustrates this estimation method.

Example 14.3: What is the oxygen concentration below which ignition of benzene cannot occur at ambient temperature? The chemical formula for benzene is C_6H_6 and its LEL is 1.3%. The value reported in Table 6.4 is 9.0% O_2 and contains a 2% safety margin. Thus, the actual concentration at which ignition can occur is 11%.

Solution

The oxidation chemistry is as follows:

$$C_6H_6 + 7\ 1/2\ O_2 \rightarrow 6\ CO_2 + 3\ H_2O$$

Since each lb-mol (or scfm) of benzene requires 7.5 lb-mol of oxygen at stoichiometric conditions, the minimum oxygen concentration below which ignition cannot occur is $7.5 \times 1.3 = 9.75\%$. This is reasonably close to the reported value and the error is toward a safer value. For VOC mixtures, the procedure for estimating the minimum oxygen concentration at which ignition can occur is similar to the method of calculating the LEL of mixtures.[15]

$$\text{Minimum oxygen concentration (\%)} = \frac{100}{(X_1/L_1 + X_2/L_2 + \ldots X_n/L_n)}$$

where
$$X_1, X_2, \ldots X_n = \text{vol\% of VOC 1,2, ...n, respectively}$$
$$L_1, L_2, \ldots L_n = \text{minimum } O_2 \text{ \% for VOC 1, 2, ... n, respectively}$$

Higher waste stream temperatures will increase the minimum oxygen concentration by approximately 8% for each 100°C (212°F) in temperature above ambient.

One strategy to prevent flashback is to mix a VOC-containing air stream with an inert (no oxygen) gas stream such that the resulting oxygen concentration is below the safe limits specified in Table 6.4 or calculated as shown above. In fact, this illustrates why a VOC is not ignitable above its UEL. The concentration of the VOC is so great as to dilute the oxygen concentration to a low level.

14.3   FLASHBACK VELOCITY

There are several techniques that can be used to prevent premature ignition of a VOC-containing waste stream that is between its LEL and UEL. One is to maintain the flow velocity of the waste stream above the flashback velocity. Flashback velocity should not be confused with fundamental burning velocity. Flashback velocities are higher. This is a consequence of the fact that the flame initially flashes back into the lower velocity gas near the wall of the pipe or duct.

The flashback velocity is not only a characteristic of a particular organic compound, but also depends on the size of the duct through which the waste gas is flowing. The flashback velocity is defined as follows:[16]

$$\text{Flashback velocity (ft/s)} = 0.2015 \times G_1 \times D$$

where

$\quad\quad G_1 \quad$ = critical boundary velocity gradient (1/s)
$\quad\quad D \quad\quad$ = duct or pipe diameter (ft)

Values of the critical velocity boundary gradient for common fuels are as follows:

Fuel	G_1 (1/s)
Methane	400
Ethane	650
Propane	600
Ethylene	1,500
Propylene	700
Hydrogen	10,000

The following example illustrates the effect of the duct diameter on the flashback velocity.

Example 14.4: Determine the flashback velocity of a 500-scfm air stream at ambient temperature containing 500 ppmv of propylene in a 4- and 6-in. duct.

Solution

$$\text{Flashback velocity (ft/s)} = 0.2015 \times G_1 \text{ (s}^{-1}) \times D \text{ (ft)}$$

	4 in. Duct	6 in. Duct
Flashback velocity (ft/s)	$0.2015 \times 700 \times 4/12 = 47.0$	$0.2015 \times 700 \times 6/12 = 70.5$
Actual velocity (ft/s)	96	43

This example indicates that flashback could occur within the 6-in. duct but not the 4-in. duct. The actual velocity through the duct should always exceed the flashback velocity plus an adequate margin of safety.

14.4   FLASHBACK PREVENTION TECHNIQUES

By maintaining the minimum flow velocity of the waste stream above the flashback velocity, flashback cannot occur. One technique to prevent flashback is to install a venturi section (reduced diameter throat) in the waste stream ductwork. The venturi throat is sized such that the velocity through the throat is above the flashback velocity at the minimum waste stream flow rate. Sometimes an inert gas such as steam or nitrogen is added to ensure this minimum velocity as shown in Figure 14.1.

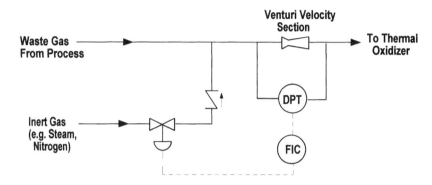

Figure 14.1   Minimum velocity technique for flashback prevention.

Another method of preventing flashback is to install a flame arrestor in the ductwork. This is a passive device that permits gas flow, but inhibits flame propagation. The most common type of flame arrestor uses a crimped sheet of steel with close, uniform spacing to provide a grid face with small openings. The elements are contained in a pressure-tight housing. This type of arrestor is illustrated in Figure 14.2.

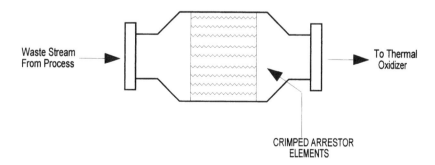

Figure 14.2   Flashback flame arrestor.

When a vapor mixture does ignite, a flame arrestor prevents further flame propagation by absorbing and dissipating heat from the burning gas or vapor on one side of the arrestor, lowering the temperature of the combustion products to below the auto-ignition point of the gas on the opposite side. This is achieved as the hot gases give up their heat to the cell walls on passing through the crimped metal arrestor elements.

A distinction must be made between a deflagration and a detonation. A **deflagration** is a subsonic flame front with explosion pressure below 300 psi. A **detonation** is a supersonic flame front that can travel at speeds up to Mach 15 and produce pressures of up to 5000 psi. When first ignited, the flame front begins as a deflagration. However, it can quickly become a detonation if aided by elbows, tees, and valves that produce turbulence. A flame arrestor chosen for a VOC application should be able to stop both types of flame fronts. Flame arrestors must be constructed to certified standards. The Coast Guard (USCG) is the agency responsible for setting such standards in the U.S.

Devices that cause turbulence have the effect of accelerating the flame. Pipe diameters can also affect flame propagation. The transition from detonation to deflagration is dependent on pipe diameter and length. Larger pipe diameters typically require longer flame lengths before reaching detonation. The distance between the flame arrestor and a potential ignition source (i.e., thermal oxidizer) should be minimized (maximum of 15 ft) and with minimal obstructions (maximum of one elbow) that could cause turbulence.

The design of the flame arrestor must correspond to the type of flammable gas that it must handle. The element must be designed to accommodate the specific gas group that could possible ignite and propagate in the system. Organic vapors and gases are grouped into four classifications, depending on their degree of difficulty of preventing flame propagation. Examples of gases and vapors in these classifications are shown in Table 14.1.

Another method of preventing flashback is to dilute the VOCs in the waste gas to below their LEL by adding dilution air. Most industrial risk insurers require that operators use LEL monitors if the VOC concentrations **of a waste stream** are expected to approach 15 to 25% of the LEL. Typically, feedback from a LEL monitor is used to control a valve that allows the addition of ambient air to dilute the organic vapors to less than 50% of the LEL. This is shown schematically in Figure 14.3. The disadvantage of this method is that this additional air increases the heat load on the thermal oxidizer and increases the auxiliary fuel required.

Installation of a seal pot in the waste gas transfer duct is yet another method of restricting flashback. The seal pot provides a function similar to the flame arrestor. While it doesn't prevent flashback from occurring, it prevents it from traveling back to the process generating the waste stream and potentially causing catastrophic damage. A schematic of a seal pot is shown in Figure 14.4. Water is normally used as the sealing fluid. Waste gas flows into the seal pot and exits the inlet pipe under the water level (typically 6 in. below water level). The waste gas is dispersed into discrete bubbles as it passes up through a perforated plate that is also below the water level. Thus, if flashback occurs in the vapor space above the water or further downstream, ignition is quenched because of the discontinuous nature (dispersed bubbles) below the water level.

TABLE 14.1
Gas and Vapor Classifications for Flame Arrestor Design

Group A

Acetylene

Group B

Butadiene
Ethylene oxide
Hydrogen (and gases containing more than 30% hydrogen)
Propylene oxide
Propyl nitrate

Group C

Acetaldehyde	Dimethyl hydrazine	Hydrogen sulfide
Cyclopropane	Ethylene	Methyl mercaptan
Diethyl ether		

Group D

Acetone	Acrylonitrile	Ammonia
Benzene	Butane	Butylene
Butanol	Cyclohexane	n-Butyl acetate
Isobutyl acetate	Ethane	Ethanol
Ethyl acetate	Ethyl acrylate	Gasoline
Ethylene dichloride	Heptanes	Hexanes
Isoprene	Methane	Methanol
Methyl acrylate	Methyl ethyl ketone	Methylamine
Methyl mercaptan	Isoamyl alcohol	Isobutyl alcohol
Methyl isobutyl ketone	tert-Butyl alcohol	Naphtha
n-Propyl acetate	Octanes	Pentanes
Amyl alcohol	Propane	Propanol
Propylene	Styrene	Toluene
Turpentine	Vinyl acetate	Vinyl chloride
Xylenes		

14.5   COMBUSTION SAFEGUARDS

The National Fire Protection Association (NFPA) among others has established standards for the design and operation of fuel-fired combustion systems. The NFPA consists of representatives from industry, utilities, insurance companies, Underwriters Laboratories Inc., Factory Mutual Research Corp., and Industrial Risk Insurers Inc. While prevention of VOC flashback and explosion are one aspect of safety that must be included in the design of a thermal oxidation system, the design and operation of the auxiliary fuel system is equally important. While not specifically oriented towards thermal oxidizers, the similarity between the combustion systems described in these standards and thermal oxidizer auxiliary

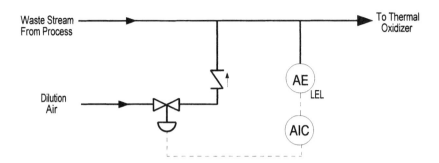

Figure 14.3 Flashback prevention by using air to dilute waste stream to below LEL.

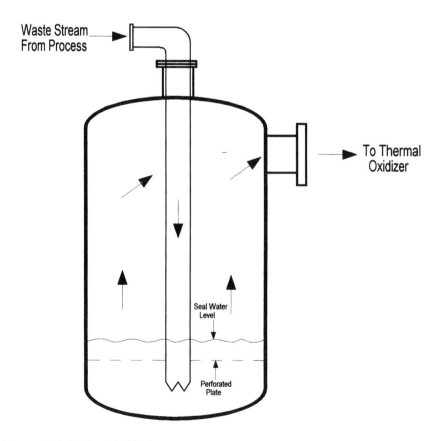

Figure 14.4 Seal pot flashback arrestor.

fuel systems makes these standards a valuable resource in the design of thermal oxidizer safety systems. The two NFPA standards most applicable to thermal oxidizer auxiliary fuel system design and operation are NPPA 8501 (Single Burner Boiler Operation) and NFPA 8502 (Standard for the Prevention of Furnace Explosions/Implosions in Multiple Burner Boilers). NPPA 8502 is a compilation of the following four prior standards:

NFPA 85B — Standard for the Prevention of Furnace Explosions in Natural Gas-Fired Multiple Burner Boiler-Furnaces
NFPA 85D — Standard for the Prevention of Furnace Explosions in Fuel Oil-Fired Multiple Burner Boiler-Furnaces
NFPA 85E — Standard for the Prevention of Furnace Explosions in Pulverized Coal-Fired Multiple Burner Boiler-Furnaces
NFPA 85G — Standard for the Prevention of Furnace Explosions in Multiple Burner Boiler-Furnaces

Some general features of these standards will be described here. However, designers and operators of VOC thermal oxidation equipment should thoroughly review these or similar standards in detail before designing, installing, or operating thermal oxidation systems.

The objective of combustion safeguard systems is to prevent the accumulation of excessive quantities of combustibles mixed with air in proportions that can result in uncontrolled combustion when an ignition source is applied. Common conditions that can produce an explosion in a thermal oxidation system are

1. Interruption of the fuel or air supply or ignition source resulting in a momentary loss of flame, followed by restoration and delayed reignition of accumulated combustible gases
2. Fuel leakage into an idle thermal oxidizer and ignition of the fuel by a spark or other ignition source
3. Repeated unsuccessful attempts to light the burner without appropriate purging between attempts, resulting in the accumulation of an explosive mixture
4. Accumulation of an explosive mixture of fuel and air as a result of a flame-out, followed by ignition of accumulated fuel and air by a spark or other ignition source

A burner management system (BMS) is a control system dedicated to safe system operation. It includes the interlock system, fuel trip system, master fuel trip system, master fuel trip relay, flame monitoring and tripping system ignition subsystem, and main burner system. The logic system provides outputs in a specific sequence in response to external inputs and internal logic. It is designed so that a single failure does not prevent a safe shutdown of the system. The BMS is also designed so that logic system failure does not prevent operator intervention.

The combustion control system should be designed to maintain the air/fuel ratio within the limits for continuous combustion and flame stability over the full operating range. The design should include provisions for setting minimum and maximum limits on the fuel and air control to prevent fuel and air flows beyond stable limits. When changing firing rates, the control system must be designed to maintain proper air/fuel ratio control.

The burner must include a flame scanner that monitors the flame, the igniter, or both. If loss of flame is detected, the fuel supply should be shut off immediately. An alarm is generally included to alert operators of a flame-out. Position sensors or interlock switches are provided to ensure the correct positioning of all dampers and valves. Provisions for visual observation (sight glass) of the burner flame should also be included.

14.6 TYPICAL NATURAL GAS FUEL TRAIN

The components of a typical natural gas fuel train are shown in Figure 14.5. The main gas line contains two safety shut-off valves. These valves are fitted with proximity switches to verify their position. A vent line and valve are included between the two shut-off valves. The vent valve is usually solenoid operated. A fuel pressure regulator is located upstream of the safety shut-off valves. A manual shut-off valve is included upstream of the pressure regulator. High and low pressure switches and alarms are located on the fuel supply line between the pressure regulator and flow control valve. Usually one or more local pressure indicators are included on the fuel supply line. The fuel flow control valve is located downstream of the second safety shut-off valve. A metering orifice is usually included downstream of the flow control valve. The combustion air line also contains a low-pressure switch and alarm.

The pilot fuel line branches upstream of the safety shut-off valves and may contain another pressure-regulating valve. A small air line from the main combustion air line feeds the pilot mixer where fuel and air are mixed before ignition (in the premixed type of pilot system). A spark igniter is installed at the end of the pilot line. A flame scanner monitors the pilot flame, main flame, or both. The flame scanner usually contains an alarm function.

14.7 START-UP SEQUENCE

The exact sequence for start-up of the system will vary depending on the system design and the specific standards followed. However, after the functionality of each item of equipment is verified, the following sequence is generally followed from a cold start:

1. Fans are started (induced draft, forced draft, or both).
2. Air valves are opened to their purge position.
3. The main safety shut-off valves (SSOV) are closed and the main fuel control valve is set to the minimum firing position.

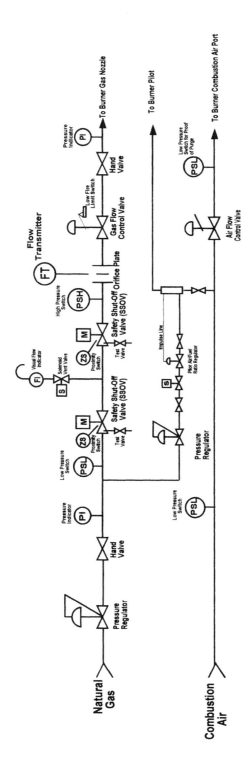

Figure 14.5 Typical natural gas fuel train.

4. The residence chamber is purged with air for four volume changes prior to lighting the burner.
5. The burner header fuel pressure is set for light-off. The burner header is vented to fill the fuel line with fuel.
6. The air flow control valve is set to the start-up position.
7. The spark to the igniter is energized to light the pilot. The pilot must light within 15 s. Otherwise, the purge cycle must be repeated.
8. Once the scanner has verified flame on the pilot, the main fuel safety shut-off valves are opened to ignite the main burner.
10. After flame on the burner is established, the air and fuel control valves are adjusted to their normal position. If ignition of the main burner is not obtained within 5 s of opening the fuel safety shut-off valves, these valves are again closed and the purge cycle repeated before attempting to relight.

Once operating at normal conditions, a repurge of the system is not required if a flame-out occurs and the burner is reignited before the combustion temperature falls below 1400°F.

14.8   INTERLOCKS

Generally, the following interlocks, as a minimum, will cause a shutdown of the burner:

Loss of flame
High fuel pressure
Low fuel pressure
Low combustion air pressure
Low combustion air flow
High chamber temperature
VOC concentration in excess of 25% of LEL (where applicable)

Interlocks are usually connected to alarms to alert the operator of the system failure.

14.9   LEAD/LAG TEMPERATURE CONTROL

A safety feature that is included in many thermal oxidizer designs is lead/lag temperature control. In this control scheme, the controls force the combustion air to *lead* the fuel flow rate in response to a decrease in temperature and a demand for additional fuel to raise the temperature. That is, the air flow rate is increased before the fuel flow rate. Conversely, if the temperature rises above the setpoint value, the fuel flow is first decreased, followed by a decrease in the air rate (to maintain the same air/fuel ratio). The air is said to *lag* the fuel in this scenario. The system's first response in all cases is to prevent excess fuel addition and a sudden rise in temperature.

14.10 ELECTRICAL HAZARD CLASSIFICATIONS

Thermal oxidation systems usually contain electric motor driven equipment, such as fans. Standards have been developed for the design, installation, and operation of this equipment in environments where an arc or thermal ignition in electrical equipment could cause ignition of flammable gas or vapor mixtures, combustible dusts, or easily ignitable fibers.

In the U.S., area classifications are given in Article 500 of the National Electric Code, ANSI/NFPA 70. Locations are classified by Class, Division, and Group. Class refers to the form of the flammable materials in the atmosphere. Division provides an indication of the probability of the presence of a flammable material in ignitable concentrations. Group is an indication of the nature of the flammable material.

Class I locations are those in which flammable gases or vapors may be present in sufficient quantities to produce explosive or ignitable mixtures. Class II locations are considered hazardous because of the presence of combustible dust. Class III atmospheres are those which contain easily ignitable fibers which are not present in sufficient quantities to produce ignitable mixtures. Class I atmospheres are the most likely to be encountered in VOC thermal oxidation applications.

Class I, Division 1 locations are those in which (1) hazardous concentrations exist under normal conditions, (2) hazardous concentrations exist because of repair or maintenance operations, and (3) breakdown or faulty operation of processes or equipment might release flammable concentrations of gases or vapors.

Class I, Division 2 locations are those in which (1) volatile flammable liquids or gases are handled and processed but are normally confined to enclosed containers and can only escape by accidental rupture or breakdown of their containers; (2) ignitable concentrations of gases or vapors are normally prevented from occurring by positive mechanical ventilation, but might become hazardous through failure or abnormal operation of the ventilating equipment; and/or (3) are areas adjacent to Division 1 locations to which ignitable concentrations of gases or vapors may be conveyed.

Groups are classified by the explosive characteristics of air mixtures of gases, vapors, or dusts and vary by the specific species involved. These species and their group classifications are the same as those used in specifying flame arrestors shown in Table 14.1. In addition, group E atmospheres are those containing combustible metal dusts or other dusts with resistivities of less than 100 ohm-cm. Group F atmospheres are those containing carbon black, charcoal, coal, or coke dusts, or dusts sensitized by other materials that have resistivities between 100 and 10^8 ohm-cm. Group G atmospheres are those containing combustible dusts with resistivities greater or equal to 10^8 ohm-cm.

Once the location of a thermal oxidizer installation has been selected, the electrical classification at that location must be determined and used for specifying the purchase of the appropriate electrical components.

15 Design Checklist

CONTENTS

Many aspects of thermal oxidizer design were discussed in previous chapters of this book. However, there are also other considerations that should not be overlooked. This chapter provides a checklist of items that should be considered in the design of a thermal oxidation system or when preparing a specification to obtain quotes on the design and purchase of this equipment.

No attempt is made to specify the exact criteria to use in formulating design standards. Different companies have different requirements with regard to quality, design standards, safety, scope of supply, etc. However, all of the items in this chapter should be at least considered. The exact requirements selected are left to the discretion of the user.

15.1 PRIMARY OBJECTIVES

When preparing an equipment specification, the purpose of the thermal oxidizer should be enumerated (e.g., to treat off-gas from XYZ production process). If the source of the waste stream is known, equipment vendors may find that they have had experience with that identical off-gas in the past. The equipment vendor should be told if minimizing auxiliary fuel consumption using heat recovery equipment is desirable. Also, if the unit includes heat recovery in the form of a waste heat boiler, is the production of steam more important or is minimizing auxiliary fuel consumption more important?

15.2   SCOPE OF SUPPLY

The end user must clearly indicate the equipment that must be included when a vendor is supplying a thermal oxidation system. This includes not only the thermal oxidizer itself, but also auxiliary equipment such as fans, instrumentation, valves, controls, ductwork, expansion joints, heat recovery equipment, sootblowers, transitions, and stacks. If a stack is included, the required height of the stack must also be specified along with the number, location, and size of emission test ports.

15.3   PROCESS CONDITIONS

The importance of specifying the complete range of process conditions is often overlooked by the end-user when specifying a thermal oxidizer. The design specifications should include not only normal operating conditions but also design rates, start-up conditions, possible upset conditions, and whether the thermal oxidizer should be maintained in a hot standby condition when the waste stream is not being generated. The temperature, pressure, and composition of the waste stream, including any particulates present, must be specified for each condition. Many times there is more than one normal operating condition.

The end user should also specify the process design criteria. Should the thermal oxidizer be designed for the maximum rate specified or is some margin above the specified rate required to account for uncertainty in the process rates and composition? Should the burner be sized based on the heat release from the volatile organic components of the waste stream, or should this heat release be ignored when sizing the burner? If process upsets can occur, resulting in a surge of VOC or waste stream flow rate, how fast does this occur? Uncertainty in answering these questions normally justifies an increased margin between the specified design conditions and the expected normal operating rates.

15.4   DESIGN REQUIREMENTS

The design specification should include the following:

- Battery limits
  - Where does the vendor's scope of supply begin?
- Forced draft, induced draft, or balanced draft system
  - Most thermal oxidizers are forced draft (i.e., positive pressure). Hazardous wastes systems are induced draft (i.e., negative pressure — leakage into system). Systems with waste heat boilers or scrubbers are sometimes balanced draft (forced draft combustion air and induced draft after scrubber or boiler)
- Orientation
  - The end user may have a preference for a vertical or horizontal orientation for reasons such as maintenance, limited footprint available, etc.
- Location
  - Indoor or outdoor

- Maximum shell skin temperature or preference for external insulation
  - Affects refractory design
  - Under what conditions (ambient temperature, wind velocity) should this be calculated?
  - Do OSHA requirements apply and to what extent?
  - External insulation raises shell metal temperature
- Shell metal thickness
  - Minimum 3/8 in. is typical
- Assembly
  - Large units may be fabricated with multiple sections
  - Are flanges acceptable or must subsections be welded together?
  - If welding is required, what weld standards apply (e.g., AWS D1.1)?
- Rain shield/exterior shroud
  - For units processing VOCs producing SO_2 or HCl, this may be necessary to maintain a hot shell and prevent dewpoint condensation and corrosion
- Sight ports
  - These can be valuable in assessing an operating problem
- Refractory temperature rating
  - The end user may want to specify that the interior refractory lining be capable of withstanding a specified temperature which is above the normal operating temperature to guard against unexpected temperature excursions
- Design codes
  - Many companies require equipment to be built to specific standards (e.g., ASTM, ANSI, ASME, AWS, etc.)
- Shell material and grade
  - Carbon steel is most common but stainless steel systems do exist
- Electrical classification
  - Class, Division, Group
- Electrical enclosures
  - NEMA rating
- Earthquake zone
- Design wind loading
- Minimum time from cold start to normal operating temperature
  - Important to match to the process that is generating the VOC-containing waste stream
- Reliability
  - On-stream factor (most thermal oxidizers can meet a 98% on-stream factor)
- Noise level
  - Maximum decibel level and at what distance from source
- Burner turndown
  - 8:1 to 10:1 is normal, but burners or systems can be designed for higher turndowns
- Waste gas turndown
  - Different from burner turndown

- Access doors
  - Number, locations, and size
- Control system
  - Local control panel, PLC, DCS, or combinations
- Instrumentation requirements
  - Redundant thermocouples in critical locations, for example
- Shop assembled vs. field erected
  - Define what is acceptable
  - Usually shop assembled unless a very large unit
  - Sometimes refractory is field installed
- Safety interlocks required
  - To meet plant standards
- Structural steel, ladders, stairways, platforms
  - Define what must be supplied
- Sandblasting/paint
  - Define requirements

15.5 PERFORMANCE REQUIREMENTS

The performance requirements of the VOC thermal oxidation system must be identified. In addition to reliability, this typically refers to emission limits and how they are measured. It is particularly important to clearly describe the acceptance criteria. For example, it is common to perform three replicate stack tests for a particular pollutant species. To satisfy a particular limit, must all three test results be below the emission limit, is two of three acceptable, or is an average of the three acceptable? When specifying VOC destruction efficiency and more than one VOC species is present in the waste stream, does this destruction efficiency apply to the average for all VOCs present or does it apply to each individual VOC?

Emission levels from a thermal oxidizer are usually measured using standard EPA test methods described in 40 CFR Part 60 of the Code of Federal Regulations. The answer to the questions above can affect the test method selected. For example, EPA Test Method 25A measures total VOC emissions, but does not distinguish between individual compounds. If that is a requirement, a method such as EPA Method 18 must be used, which measures the concentration of individual organic compounds. However, EPA Method 18 requires special analyzers and will be more expensive. The following species are most commonly of concern in performing analyses of stack gases from a thermal oxidizer:

VOC destruction efficiency (DRE)
Nitrogen oxides (NOx)
Carbon monoxide (CO)
Sulfur oxides (SOx)
Hydrogen chloride (HCl)
Particulates (size range may also be specified [e.g., PM-2.5, particulate less
   than 2.5 microns])

In addition to emissions performance, an auxiliary fuel consumption guarantee is sometimes included, particularly if the system includes heat recovery.

The thermal oxidation system is usually subjected to a performance test during start-up. This test is intended to verify operability, fuel consumption, and compliance with emission limits. Equally important to the tests themselves is a clear definition of the consequences, should the initial tests fail to meet the requirements. Typically, the equipment vendor is given a certain period of time to modify the equipment before retesting. Responsibility for the cost of the second (or subsequent tests) should also be stated in the equipment specification.

15.6 AUXILIARY EQUIPMENT

A thermal oxidation system requires ancillary equipment in addition to the thermal oxidizer itself. Items to consider when specifying this equipment are

- Fans
  - Many times the motor voltage used is related to the fan horsepower
  - Motor service factor
  - Motor thermal overload protection
  - Maximum speed
  - Discharge orientation
  - Fan arrangement
  - Method of flow control (e.g., inlet vane damper or variable speed drive)
  - Hazard classification (e.g., Class I, Division II)
  - Shaft sealing method
  - Material of construction
  - AMCA rating
  - Vibration monitors
  - Drive (direct, belt, etc.)
  - Rating (flow at specified pressure)
- Heat exchangers
  - Inlet temperature or range
  - Performance required (e.g., preheat temperature)
  - Plate or shell and tube
  - Vibration analysis
  - Materials of construction (alternately, specify gas composition and allow vendor to select materials)
  - Minimum casing thickness
  - Flow rating
  - Pressure rating
  - Maximum footprint available
  - Inlet/outlet orientation (to match thermal oxidizer)
- Stack
  - Inlet temperature or range
  - Flue gas composition
  - Free-standing or guyed

- Materials of construction (e.g., FRP, carbon steel)
- Height
- Number of flues
- Number of test ports, size, and elevation
- Platform for testing (also monorail for inserting test probes)
- Electrical receptacles on platform
- Lighting
- Ladders and cages
- Rain cap
- Design wind loading (see ANSI A58.1)
- Seismic loading
- Maximum flue gas velocity
- Insulation (external, internal, type)
- Aircraft warning lights
- Surface preparation (SSPC standard)/painting
- Access door (number, location, size)
- Continuous emission monitoring system (CEMS)
- Waste heat boiler
  - Components required (e.g., screen, superheater, evaporator, economizer)
  - Firetube or water tube
  - POC flow rate
  - Inlet gas temperature and composition
  - Boiler exit gas temperature
  - Boiler tube maximum fin density
  - Fouling factor
  - Design steam temperature/pressure
  - Steam purity
  - Maximum superheated steam temperature variation (important if steam used in turbines)
  - Drum water retention time
  - Feedwater conditions (temperature and purity)
  - Battery limits
  - ASME Code design
  - Sootblowers (number, location, permanent insertion or retractable, etc.)
  - Expansion joints
  - Access doors (size and locations)
  - Ash hopper
  - Boiler trim
  - Insulation (internal, external type)
  - Surface preparation (SSPC standard)/painting
  - Orientation (to match thermal oxidizer)
  - Ladders, platforms, walkways, stairways, grating
  - Shop fabricated vs. field-erected components

15.7   UTILITIES AVAILABLE

When requesting a quote for thermal oxidation equipment from vendors, the utilities available in the plant should be specified.

- Natural gas
  - Heating value or composition, specific gravity, pressure
- Fuel oil
  - Grade, heating value, purity, viscosity
- Electrical power
  - Voltage, phase, frequency
- Instrument air
  - Pressure, temperature, dew point
- Cooling water
  - Pressure, temperature, purity
- Steam for atomization
  - Pressure, condition (saturated, degrees of superheat)
- Boiler feedwater
  - Temperature, pressure, purity
- Air for atomization
  - Pressure, temperature, dew point

15.8   ENVIRONMENT

- Site elevation
- Summer design temperatures
  - Dry bulb
  - Wet bulb
- Winter design temperature
- Design rainfall
- Design wind velocity
- Seismic zone

15.9   PREFERRED EQUIPMENT/APPROVED VENDORS

Many times, a plant will try to limit auxiliary equipment, instrumentation, and controls to a certain group of vendors. This is done because operating personnel are familiar with the operation of this equipment, and also to reduce the number of spares required. If this is the case, the preferred vendor list should be included in the thermal oxidizer equipment specification.

15.10   START-UP ASSISTANCE

When obtaining a quote from a thermal oxidizer vendor, the costs for start-up assistance should be clearly delineated in the vendor proposal. Sometimes a certain amount of time is included free of charge as part of the equipment order. However, problems develop many times when the start-up is unexpectedly prolonged and each party expects the other to be responsible for the time over and above that included with the equipment order.

Training of plant personnel is usually conducted during the start-up period. Again, the responsibility for these costs must be clearly defined.

15.11   SPARE PARTS

A recommended spare parts list and cost is usually requested.

15.12   DESIGN DOCUMENTATION

The request for a quote from a thermal oxidizer equipment vendor should indicate the extent of documentation that will be required if an order is placed. The following should be considered:

- General arrangement drawings
- Refractory cross section drawings
- Piping and instrumentation drawings (P&ID)
- "As-built" drawings (when available)
- Foundation and loading diagrams
- Process flow diagrams
- Electrical drawings
- Control logic diagrams
- ISA data sheets
- Structural drawings
- Dimensioned front and side sectional elevation drawings of all major equipment
- Outline drawings showing the principal dimensions, methods of support, weights to be supported, and locations of weights
- Fan performance curves
- Noise level data sheets for fans
- Any deviations from the specification
- Schedule for drawings, equipment fabrication, and installation
- Operating instructions (and number of copies required)
- Maintenance instructions
- Catalog cut sheets for all equipment
- Burner curves showing pressure vs. firing rate

Appendix A

Incinerability Ranking

Class 1

Hydrogen cyanide
Cyanogen
Benzene
Sulfur hexafluoride
Napthalene
Fluoranthene
Benzo(*j*)fluoranthene
Benzo(*b*)fluoranthene
Benzanthracene
Chrysene
Benzo(*a*)pyrene
Dibenz(*a,h*)anthracene
Indeno(*1,2,3-cd*)pyrene
Dibenzo(*a,h*)pyrene
Dibenzo(*a,e*)pyrene

Cyanogen chloride
Acetonitrile
Chlorobenzene
Acrylonitrile
Dichlorobenzene
Trichlorobenzene
Tetrachlorobenzene
Methyl chloride
Tetrachlorobenzene
Pentachlorobenzene
Hexachlorobenzene
Methyl bromide
Tetrachlorodibenzo-*p*-dioxin
Dibenzo(*a,i*)pyrene

Class 2

Toluene
Chloroaniline
1,1 Dichloro-2,2-*bis*-(4-chlorophenylethylene)
Trichloroethene
Dichloroethene
Dimethylbenz(*a*)anthracene
Formaldehyde
Methyl chlorocarbonate
Tetrachlorodibenzo-*p*-dioxin
Malononitrile
Dichloroethene
Acrylamide
Dichloromethane
Chloroaniline
Chloro-1,3-butadiene
Pronamide
Crotonaldehyde
Chlorocresol
Adrenaline

Tetrachloroethane
Formic acid
Phosgene

Diphenylamine
Fluoroacetic acid
Aniline
Malonitrile
Methyl isocyanate
Dimethylphenethylamine
Naphthylamine
Fluoroacetamide
Methyl methacrylate
Methacrylonitrile
Methylcholanthrene
Diphenylamine
Acetylaminofluorene
Dichlorophenol
Methylactonitrile
Dimethylphenol

continued

Incinerability Ranking (continued)

Class 3

Aminobiphenyl	Dichlorobenzidine
Chlorophenol	Benzidine
Dimethylbenzadine	Phenylenediamine
n-Propylamine	Pyridine
Chlorophenol	Picoline
Dichloropropene	Thioacetamide
Trichlorotrifluoroethane	Phenol
Benz(c)acridine	Dichlorodifluoromethane
Acetophenone	Trichlorofluoromethane
Trichlorofluoromethane	Propionitrile
Benzoquinone	Vinyl chloride
Dibenz(a,h)acridine	Dibenz(a,j)acridine
Hexachlorobutadiene	Naphthoquinone
Dimethyl phthalate	Acetyl chloride
Acetonylbenzyl-4-hydroxycoumarin	Maleic anhydride
Chlorophenol	Dichloro-2-butene
Dichloropropene	Dibenzo(c,g)carbazole
Toluenediamine	Resorcinol
Cresol	Dichlorophenol
Methyl ethyl ketone	Diethylstilbesterol
Benzenethiol	Isobutyl alcohol

Class 4

Chloropropene	Chloropropionitrile
Dichloropropene	Dichloro-2-propanol
Tetrachloroethane	Dichlorodiphenyldichloroethane
Trichlorophenol	Dichloro-2-propanol
Ethyl chloride	Phthalic anhydride
Dichloropropene	Methyl parathion
Hydrazine	Nitrophenol
Benzyl chloride	Chlorodifluoromethane
Dibromomethane	Pentachlorophenol
Dichloroethane	Hexachlorocyclohexane
Mustard gas	Dichlorofluoromethane
Nitrogen mustard	Dintrobenzene
Chlornaphazine	Nitroaniline
Dichloropropene	Pentachloroethane
Dichloro-2-butene	Dinitrobenzene
Tetrachlorophenol	Trichloroethane
Tetrachloromethane	Trichloromethane
Bromoacetone	Dieldrin
Hexachlorophene	Isodrin
1,4 Dioxane	Aldrin
Chloroambucil	Dichlropropane
Nitrobenzene	Nitrotoluidine

Incinerability Ranking (continued)

Class 4 (continued)

Chloroacetaldehyde
Dinitrotoluene
Benzal chloride
Ethylene oxide
Dimethylcarbamoylchloride
Dichlorodiphenyltrichloroethane
Auramine
Dichloropropane
Dinitrophenol
Trinitrobenzene
Cyclohexyl-4,6-dinitrophenol
Chloral
Dinitrocresol
Diepoxybutane

Trichloropropnae
Hexachlorocyclopentadiene
Dichloro-1-propanol
Dichloroethane
Glycidaldehyde
Dichloropropane
Heptachlor
Chloro-2,3-epoxypropane
Bis(2-chloroethyl)ether
Butyl-4,6-dinitrophenol
Bis(2-chloroethoxy)methane
Trichloromethanethiol
Heptachlor epoxide

Class 5

Benzotrichloride
Methapyrilene
Phenacetin
Methyl hydrazine
Dibromoethane
Alfatoxins
Trichloroethane
Hexachloroethane
Bromoform
Chlorobenzilate
Ethyl carbamate
Ethyl methacrylate
Lasiocarpine
Amitrole
Muscimol
Iodomethane
Dichlorophenoxyacetic acid
Chloroethylvinylether
Methylene bis(2-chloroaniline)
Dibromo-3-chloropropane
Tetrachloroethane
Dimethylhydrazine
Methyl-2-methylthiopropionaldehyde-*O*-
  (methylcarbonyl)oxime
Trichlorophenoxypropionic acid
Methylazirdine
Brucine
Isosafrole

Chloromethylmethyl ether
Thiofanox
Dimethylhydrazine
Chlordane
bis(chloromethyl)ether
Parathion
Dichloropropane
Maleic hydrazide
Bromophenyl phenyl ether
Bis(2chloroisopropyl)ether
Dihydrosafrole
Methyl methanesulfonate
Propane sulfone
Saccharin
N,N-Diethylhydrazine
Methyomyl
Hexachloropropene
Pentachloronitrobenzene
Diallate
Ethyleneimine
Aramite
Dimethoate
Trichlorophenoxyacetic acid

Tris(2,3-dibromopropyl)phosphate
Methoxychlor
Kepone
Safrole

continued

Incinerability Ranking (continued)

Class 5 (continued)

Tris(1-azridinyl) phosphine sulfide

O,O-Diethylphosphoric acid,*O*-*p*-nitrophenyl ester

Dimethoxybenzidine

Diphenylhydrazine

Class 6

n-Butylbenzyl phthalate

O,O-Diethyl-*O*-pyrazinyl phosphorothioate

Dimethylaminoazobenzene

Diethyl phthalate

O,O-Diethyl-S-methyl ester of phosphoric acid

Citrus Red No. 2

Trypan blue

Ethyl methanesulfonate

Disulfoton

Diisopropylfluorophosphate

O,O,O-Triethyl phosphorothioate

Di-*n*-butyl phthalate

Octamethylpyrophosphoramide

Bis(2-ethylhexyl)phthalate

Methylthiouracil

Propylthiouracil

O,O-Diethyl-*S*-{(ethylthio)methyl}ester of phosphorodithioic acid

Class 7

Strychnine

Cyclophosphamide

Nicotine

Reserpine

Toluidine hydrochloride

Tolylene diisocyanate

Endrin

Butanone peroxide

Tetraethylpyrophosphate

Nitroglycerine

Tetraethyldithiopyrophosphate

Ethylene-bis-dithiocarbamic acid

Tetranitromethane

Uracil mustard

Acetyl-2-thiourea

Chlorophenyl thiourea

N-Phenylthiourea

Naphthyl-2-thiourea

Thiourea

Daunomycin

Ethylene thiourea

Thiosemicarbazide

Melphalan

Di-*n*-propylnitrosamine

Endosulfan

Dithiobiuret

Thiuram

Azaserine

Hexaethyl tetraphosphate

Nitrogen mustard *N*-oxide

Nitroquinoline-1-oxide

Cycasin

Streptozotocin

N-Methyl-*N*-nitro-*N*-nitrosoguanidine

N-Nitroso-diethanolamine

N-Nitroso-dibutylamine

N-Nitroso-*N*-ethylurea

N-Nitroso-*N*-methylurea

N-Nitroso-*N*-methylurethane

N-Nitrosodiethylamine

N-Nitrosodimethylamine

N-Nitrosomethylethylamine

N-Nitrosomethylvinylamine

N-Nitrosomorpholine

N-Nitrosonornicotine

N-Nitrosopiperidine

N-Nitrososarcosine

Nitrosopyrrolidine

Endothal

Appendix B

Table of the Elements

Element	Symbol	Atomic Number	Atomic Weight
Actinium	Ac	89	227
Aluminum	Al	13	26.982
Americium	Am	95	243
Antimony	Sb	51	121.75
Argon	Ar	18	39.948
Arsenic	As	33	74.923
Astatine	At	85	210
Barium	Ba	56	137.34
Berkelium	Bk	97	249
Berylium	Be	4	9.012
Bismuth	Bi	35	79.904
Boron	B	5	10.811
Bromine	Br	35	79.904
Cadmium	Cd	48	112.40
Calcium	Ca	20	40.08
Californium	Cf	98	251
Carbon	C	6	12.011
Cerium	Ce	58	140.12
Cesium	Cs	55	132.905
Chlorine	Cl	17	35.453
Chromium	Cr	24	51.996
Cobalt	Co	27	58.933
Copper	Cu	29	63.546
Curium	Cm	96	247
Dysprosium	Dy	66	162.50
Einsteinium	Es	99	254
Erbium	Er	68	167.26
Europium	Eu	63	151.96
Fermium	Fm	100	253
Fluorine	F	9	18.998
Francium	Fr	87	223
Gadolinium	Gd	64	157.25
Gallium	Ga	31	69.72
Germanium	Ge	32	72.59
Gold	Au	79	196.967
Hafnium	Hf	72	178.49
Helium	He	2	4.003
Holmium	Ho	67	164.93

continued

Table of the Elements (continued)

Element	Symbol	Atomic Number	Weight
Hydrogen	H	1	1.008
Indium	In	49	114.82
Iodine	I	53	126.904
Iridium	Ir	77	192.2
Iron	Fe	26	55.847
Krypton	Kr	36	83.80
Lanthanum	La	57	138.91
Lawrencium	Lw	103	257
Lead	Pb	82	207.19
Lithium	Li	3	6.939
Lutetium	Lu	71	174.97
Magnesium	Mg	12	24.312
Manganese	Mn	25	54.938
Mendelevium	Md	101	256
Mercury	Hg	80	200.59
Molybdenum	Mo	42	95.94
Neodymium	Nd	60	144.24
Neon	Ne	10	20.183
Neptunium	Np	93	237
Nickel	Ni	28	58.71
Niobium	Nb	41	92.906
Nitrogen	N	7	14.007
Nobelium	No	102	254
Osmium	Os	76	190.2
Oxygen	O	8	15.999
Palladium	Pd	46	106.4
Phosphorus	P	15	30.974
Platinum	Pt	78	195.09
Plutonium	Pu	94	242
Polonium	Po	84	210
Potassium	K	49	39.102
Praseodymium	Pr	59	140.907
Promethium	Pm	61	145
Protactinium	Pa	91	231
Radium	Ra	88	226
Radon	Rn	86	222
Rhenium	Re	75	186.2
Rhodium	Rh	45	102.905
Rubidium	Rb	37	85.47
Ruthenium	Ru	44	101.07
Samarium	Sm	62	150.35
Scandium	Sc	21	44.956
Selenium	Se	34	78.96
Silicon	Si	14	28.086

Table of the Elements (continued)

Element	Symbol	Atomic Number	Atomic Weight
Silver	Ag	47	107.868
Sodium	Na	11	22.99
Strontium	Sr	38	87.62
Sulfur	S	16	32.064
Tantalum	Ta	73	180.948
Technetium	Tc	43	99
Tellurium	Te	52	127.60
Terbium	Tb	65	158.924
Thallium	Tl	81	204.37
Thorium	Th	90	232.038
Thulium	Tm	69	168.934
Tin	Sn	50	118.69
Titanium	Ti	22	47.90
Tungsten	W	74	183.85
Uranium	U	92	238.03
Vanadium	V	23	50.942
Xenon	Xe	54	131.30
Ytterbium	Yb	70	173.04
Yttrium	Y	39	88.905
Zinc	Zn	30	65.37
Zirconium	Zr	40	91.22

Appendix C

Heats of Combustion of Organic Compounds
(All Compounds in the Vapor Phase)

Compound	Chemical Formula	Heat of Combustion (Btu/lb - LHV)
Acetaldehyde	C_2H_4O	10,854
Acetic acid	$C_2H_4O_2$	5,663
Acetic anhydride	$C_4H_6O_3$	7,280
Acetone	C_3H_6O	12,593
Acetonitrile	C_2H_3N	12,940
Acetylene	C_2H_2	20,776
Acrolein	C_3H_4O	12,741
Acrylic acid	$C_3H_4O_2$	7,969
Acrylonitrile	C_3H_3N	14,565
Ammonia	NH_3	7,992
Amyl acetate	$C_7H_{14}O_2$	13,614
Amyl alcohol	$C_5H_{12}O$	16,417
Amyl chloride	$C_5H_{11}Cl$	13,707
Amylene	C_5H_{10}	19,363
Aniline	C_6H_7N	15,246
Benzene	C_6H_6	17,446
Benzyl chloride	C_7H_7Cl	12,251
Bromobenzene	C_6H_5Br	8,559
Butadiene	C_4H_6	19,697
Butane	C_4H_{10}	19,680
Butanol	$C_4H_{10}O$	14,486
Butene	C_4H_8	19,517
Butyl acetate	$C_6H_{12}O_2$	12,360
Butyl acrylate	$C_{11}H_{20}O_2$	14,678
Butyl amine	$C_4H_{11}N$	17,812
Butyl carbitol	$C_8H_{18}O_3$	11,030
Butyl cellosolve	$C_6H_{14}O_2$	7,408
Butyl cellosolve acetate	$C_8H_{16}O_3$	14,120
Carbitol	$C_6H_{14}O_3$	11,540
Carbon disulfide	CS_2	6,231
Carbon monoxide	CO	4,347
Carbonyl sulfide	COS	3,940
Cellosolve	$C_4H_{10}O_2$	13,191
Cellosolve acetate	$C_6H_{12}O_3$	10,948
Chlorobenzene	C_6H_5Cl	11,772
Chloroform	$CHCl_3$	1,836

continued

Heats of Combustion of Organic Compounds
(All Compounds in the Vapor Phase) (continued)

Compound	Chemical Formula	Heat of Combustion (Btu/lb - LHV)
Chloroprene	C_4H_5Cl	10,922
Cumene	C_9H_{12}	17,873
Cyanogen	C_2N_2	9,053
Cyclohexane	C_6H_{12}	18,818
Dichloroethane	$C_2H_4Cl_2$	4,906
Dichloroethylene	$C_2H_2Cl_2$	4,990
Diethyl ether	$C_4H_{10}O$	14,788
Diethylamine	$C_4H_{11}N$	18,188
Dimethyl acetamide	C_4H_9O	13,984
Dimethylamine	C_2H_7N	16,800
Dimethyl disulfide	$C_2H_6S_2$	9,624
Dimethyl ether	C_2H_6O	13,450
Dimethyl formamide	C_3H_7NO	11,528
Dimethyl sulfide	C_2H_6S	13,394
Dioxane	$C_4H_8O_2$	11,768
Ethane	C_2H_6	20,432
Ethanol	C_2H_6O	12,022
Ethyl acetate	C_4H_8O	10,390
Ethyl acrylate	$C_5H_8O_2$	11,978
Ethylamine	C_2H_7N	16,433
Ethylbenzene	C_8H_{10}	17,779
Ethyl chloride	C_2H_5Cl	8,793
Ethyl mercaptan	C_2H_6S	12,399
Ethylene	C_3H_4	20,295
Ethylene dichloride	$C_2H_4Cl_2$	5,221
Ethylene glycol	$C_2H_6O_2$	7,758
Ethylene oxide	C_2H_4O	11,729
Ethyleneimine	C_2H_5N	16,291
Ethylene glycol ethyl ether acetate	$C_6H_{12}O_3$	10,948
Formaldehyde	CH_2O	7,603
Formic acid	CH_2O_2	2,481
Furfural	$C_5H_4O_2$	10,681
Heptane	C_7H_{16}	19,443
Hexane	C_6H_{14}	19,468
Hexyl cellosolve	$C_8H_{18}O_2$	7,724
Hydrogen	H_2	51,623
Hydrogen cyanide	HCN	11,004
Hydrogen sulfide	H_2S	6,545
Isobutyl alcohol	$C_4H_{10}O$	14,468
Isopropyl acetate	$C_5H_{10}O_2$	9,570
Isopropyl benzene	C_9H_{12}	17,873
Maleic anhydride	$C_4H_2O_3$	5,903

Heats of Combustion of Organic Compounds
(All Compounds in the Vapor Phase) (continued)

Compound	Chemical Formula	Heat of Combustion (Btu/lb - LHV)
Methane	CH_4	21,520
Methanol	CH_4O	91,68
Methyl acetate	$C_3H_6O_2$	9,434
Methyl amyl ketone	$C_7H_{14}O$	13,928
Methyl bromide	CH_3Br	3,188
Methyl carbitoal	$C_5H_{12}O_3$	10,990
Methyl cellosolve	$C_3H_8O_2$	9,683
Methyl chloride	CH_3Cl	6,388
Methyl cyclopentane	C_6H_{12}	18,930
Methyl ethyl ketone	C_4H_8O	13,671
Methyl formate	$C_5H_8O_2$	5,852
Methyl isobutyl ketone	$C_6H_{12}O$	12,373
Methyl mercaptan	CH_4S	10,449
Methyl methacrylate	$C_5H_8O_2$	11,177
Methyl pentane	C_6H_{14}	18,917
Methyl propyl ketone	$C_5H_{10}O$	14,466
Methyl pyrrolidone	C_5H_9NO	13,000
Methylene chloride	CH_2Cl_2	2,264
Monomethylamine	CH_5N	13,640
Napthalene	$C_{10}H_8$	16,708
Nitromethane	CH_3NO_2	4,841
Octane	C_8H_{18}	19,227
Octyl acetate	$C_{11}H_{22}O_2$	11,361
Oxo-octyl acetate	$C_{11}H_{22}O_2$	11,361
Pentane	C_5H_{12}	19,517
Pentene	C_5H_{10}	19,363
Phenol	C_6H_6O	13,688
Phosphine	PH_3	14,237
Propadiene	C_3H_4	19,634
Propane	C_3H_8	19,944
Propanol	C_3H_8O	13,652
Propionaldehyde	C_3H_6O	12,681
Propylene	C_3H_6	19,691
Propylene glycol	$C_3H_8O_2$	9,581
Propylene oxide	C_3H_6O	12,995
Pyridine	C_5H_5N	14,583
Styrene	C_8H_8	17,664
Tetrahydrofuran	C_4H_8O	15,170
Toluene	C_7H_8	17,681
Trichloroethane	$C_2H_3Cl_3$	3,682

continued

Heats of Combustion of Organic Compounds
(All Compounds in the Vapor Phase) (continued)

Compound	Chemical Formula	Heat of Combustion (Btu/lb - LHV)
Trichloroethylene	C_2HCl_3	3,235
Triethylamine	$C_6H_{15}N$	8,276
Vinyl acetate	$C_4H_6O_2$	9,960
Vinyl chloride	C_2H_3Cl	8,136
Xylene	C_8H_{10}	17,760

Appendix D

Abbreviated Steam Tables

Steam Pressure (psia)	Saturation Temperature (°F)	Steam Enthalpy (Btu/lb)							
		Sat.	350°F	400°F	450°F	500(F	600°F	700°F	800°F
50	281	1174.1	1209.9	1234.9	1259.6	1284.1	1332.9	1382.0	1431.7
100	328	1187.2	1199.9	1227.4	1253.7	1279.3	1329.6	1379.5	1429.7
150	358	1194.1	–	1219.1	1247.4	1274.3	1326.1	1376.9	1427.6
200	382	1198.3	–	1210.1	1240.6	1269.0	1322.6	1374.3	1425.5
250	401	1201.1	–	–	1233.4	1263.5	1319.0	1371.6	1423.4
300	417	1202.9	–	–	1225.7	1257.7	1315.2	1368.9	1421.3
350	432	1204.0	–	–	1217.5	1251.5	1311.4	1366.2	1419.2
400	445	1204.6	–	–	1208.8	1245.1	1307.4	1363.4	1417.0
500	467	1204.7	–	–	–	1231.2	1299.1	1357.7	1412.7
550	477	1204.3	–	–	–	1223.7	1294.3	1354.0	1409.9
600	486	1203.7	–	–	–	1215.9	1290.3	1351.8	1408.3
650	495	1202.8	–	–	–	1207.6	1285.7	1348.7	1406.0
700	503	1201.8	–	–	–	–	1281.0	1345.6	1403.7
750	511	1200.7	–	–	–	–	1276.1	1342.5	1401.5
800	518	1199.4	–	–	–	–	1271.1	1339.3	1399.1

Appendix E

Explosive Limits of Volatile Organic Compounds

Compound	Formula	LEL (%)	UEL (%)
Acetaldehyde	C_2H_4O	3.97	57.00
Acetic acid	$C_2H_4O_2$	5.40	16.00
Acetic anhydride	$C_4H_6O_3$	2.70	10.00
Acetone	C_3H_6O	2.55	12.80
Acetonitrile	C_2H_3N	4.40	16.00
Acetylene	C_2H_2	2.50	80.00
Acrolein	C_3H_4O	2.80	31.00
Acrylic acid	$C_3H_4O_2$	2.40	–
Acrylonitrile	C_3H_3N	3.05	17.00
Ammonia	NH_3	15.50	27.00
Amyl acetate	$C_7H_{14}O_2$	1.10	7.50
Amyl alcohol	$C_5H_{12}O$	1.20	9.00
Amyl chloride	$C_5H_{11}Cl$	1.60	8.60
Amylene	C_5H_{10}	1.42	8.70
Aniline	C_6H_7N	1.30	11.00
Benzene	C_6H_6	1.40	7.10
Benzyl chloride	C_7H_7Cl	1.10	–
Bromobenzene	C_6H_5Br	1.60	–
Butadiene	C_4H_6	2.00	11.50
Butane	C_4H_{10}	1.86	8.41
Butanol	$C_4H_{10}O$	1.40	11.20
Butene	C_4H_8	1.65	9.95
Butyl acetate	$C_6H_{12}O_2$	1.70	7.60
Butyl acrylate	$C_{11}H_{20}O_2$	1.40	9.40
Butyl amine	$C_4H_{11}N$	1.70	8.90
Butyl cellosolve	$C_6H_{14}O_2$	1.10	10.60
Butyl cellosolve acetate	$C_8H_{16}O_3$	0.90	8.50
Carbitol	$C_6H_{14}O_3$	1.20	8.50
Carbon disulfide	CS_2	1.30	50.00
Carbon monoxide	CO	12.00	75.00
Carbonyl sulfide	COS	11.90	28.50
Cellosolve	$C_4H_{10}O_2$	1.80	14.00
Cellosolve acetate	$C_6H_{12}O_3$	1.70	6.70
Chlorobenzene	C_6H_5Cl	1.30	7.10
Chloroprene	C_4H_5Cl	4.00	20.00
Cumene	C_9H_{12}	0.88	6.50
Cyanogen	C_2N_2	6.60	43.00
Cyclohexane	C_6H_{12}	1.26	7.75
Dichloroethane	$C_2H_4Cl_2$	5.60	11.40

continued

Explosive Limits of Volatile Organic Compounds (continued)

Compound	Formula	LEL (%)	UEL (%)
Dichloroethylene	$C_2H_2Cl_2$	9.70	12.80
Diethyl ether	$C_4H_{10}O$	1.90	36.00
Diethylamine	$C_4H_{11}N$	1.80	10.10
Dimethyl acetamide	C_4H_9O	2.80	14.40
Dimethylamine	C_2H_7N	2.80	14.40
Dimethyl disulfide	$C_2H_6S_2$	2.20	20.00
Dimethyl ether	C_2H_6O	2.00	50.00
Dimethyl formamide	C_3H_7NO	2.20	15.20
Dimethyl sulfide	C_2H_6S	2.20	19.70
Dioxane	$C_4H_8O_2$	1.97	22.50
Ethane	C_2H_6	3.00	12.50
Ethanol	C_2H_6O	3.30	19.00
Ethyl acetate	C_4H_8O	2.20	9.00
Ethyl acrylate	$C_5H_8O_2$	1.80	9.50
Ethylamine	C_2H_7N	3.50	14.00
Ethylbenzene	C_8H_{10}	1.00	–
Ethyl chloride	C_2H_5Cl	4.00	14.80
Ethyl mercaptan	C_2H_6S	2.80	18.00
Ethylene	C_3H_4	2.75	28.60
Ethylene dichloride	$C_2H_4Cl_2$	6.20	15.90
Ethylene glycol	$C_2H_6O_2$	3.20	–
Ethylene oxide	C_2H_4O	3.00	100.0
Ethyleneimine	C_2H_5N	3.30	54.80
Ethylene glycol ethyl ether acetate	$C_6H_{12}O_3$	1.70	6.70
Formaldehyde	CH_2O	7.00	73.00
Formic acid	CH_2O_2	18.00	57.00
Furfural	$C_5H_4O_2$	2.10	19.30
Heptane	C_7H_{16}	1.10	6.70
Hexane	C_6H_{14}	1.18	7.40
Hydrogen	H_2	4.00	75.00
Hydrogen cyanide	HCN	5.60	40.00
Hydrogen sulfide	H_2S	4.30	45.00
Isobutyl alcohol	$C_4H_{10}O$	1.60	10.90
Isopropyl acetate	$C_5H_{10}O_2$	1.80	8.00
Isopropyl alcohol	C_3H_8O	2.30	12.70
Isopropyl benzene	C_9H_{12}	0.88	6.50
Maleic anhydride	$C_4H_2O_3$	1.40	7.10
Methane	CH_4	5.00	15.00
Methanol	CH_4O	6.72	36.50
Methyl acetate	$C_3H_6O_2$	3.10	16.00
Methyl bromide	CH_3Br	10.00	15.00
Methyl carbital	$C_5H_{12}O_3$	1.20	–
Methyl cellosolve	$C_3H_8O_2$	2.50	19.80
Methyl chloride	CH_3Cl	8.10	17.20
Methyl cyclopentane	C_6H_{12}	1.10	8.70

Explosive Limits of Volatile Organic Compounds (continued)

Compound	Formula	LEL (%)	UEL (%)
Methyl ethyl ketone	C_4H_8O	1.80	11.50
Methyl formate	$C_5H_8O_2$	5.00	22.70
Methyl isobutyl ketone	$C_6H_{12}O$	1.40	7.50
Methyl mercaptan	CH_4S	3.90	21.80
Methyl methacrylate	$C_5H_8O_2$	2.10	12.50
Methyl pentane	C_6H_{14}	1.20	–
Methyl propyl ketone	$C_5H_{10}O$	1.10	9.65
Methylene chloride	CH_2Cl_2	12.00	19.00
Monomethylamine	CH_5N	4.95	20.75
Napthalene	$C_{10}H_8$	0.90	5.90
Nitromethane	CH_3NO_2	7.30	–
Octane	C_8H_{18}	0.95	–
Octyl acetate	$C_{11}H_{22}O_2$	0.76	8.14
Pentane	C_5H_{12}	1.40	7.80
Pentene	C_5H_{10}	1.42	8.70
Phenol	C_6H_6O	1.70	8.60
Phosphine	PH_3	1.60	98.00
Propadiene	C_3H_4	2.60	–
Propane	C_3H_8	2.12	9.35
Propanol	C_3H_8O	2.10	13.50
Propionaldehyde	C_3H_6O	2.60	16.10
Propylene	C_3H_6	2.00	11.10
Propylene gycol	$C_3H_8O_2$	2.60	12.50
Propylene oxide	C_3H_6O	2.10	38.50
Pyridine	C_5H_5N	1.80	12.40
Styrene	C_8H_8	1.10	6.10
Tetrahydrofuran	C_4H_8O	1.80	11.80
Toluene	C_7H_8	1.27	7.00
Trichloroethane	$C_2H_3Cl_3$	7.00	16.00
Trichloroethylene	C_2HCl_3	8.00	10.50
Triethylamine	$C_6H_{15}N$	1.20	8.00
Vinyl acetate	$C_4H_6O_2$	2.60	13.40
Vinyl chloride	C_2H_3Cl	4.00	26.00
Xylene	C_8H_{10}	1.00	6.00

References

1. *Chemical Week*, September 18, 1996, p. 64.
2. Morgan, J.L., Hansen, G.M., Whipple, N., and Lee, K.C., Revised model for the prediction of the time-temperature requirements for thermal destruction of dilute organic vapors and its usage for predicting compound destructability, presented at 75th Annu. Meet. Air Pollution Control Assoc., New Orleans, June 20 to 25, 1982.
3. Dellinger, B.D., Taylor, P.H., and Lee, C.C., Development of thermal stability ranking of hazardous organic compound incinerability, *Environ. Sci. Tech.*, vol. 24, March 1990.
4. Reynolds, J.P., Dupont, R.R., and Theordore, L., *Hazardous Waste Incineration Calculations — Problems and Software*, John Wiley & Sons, New York, 1991.
5. Chang, Y.C., *Pollution Eng.*, vol. 14, 1982.
6. Robinson, R.N., *Chemical Engineering Reference Manual*, 4th ed., Professional Publications, Belmont, CA, 1987.
7. Walas, S.M., *Chemical Process Equipment — Selection and Design*, Butterworths, Stoneham, MA, 1988.
8. Patrick, M.A., Experimental investigation of mixing and flow in a round, turbulent jet injected perpendicularly into a main stream, *Transa. Inst. Chem. Eng.*, vol. 45, 1967.
9. Ganapathy, V.G., *Waste Heat Boiler Deskbook*, Prentice Hall, Englewood Cliffs, NJ, 1991.
10. Stoa, T.A., Formulas estimate data for dry saturated steam, *Chem. Eng.*, December 10, 1984.
11. Ganapathy, V.G., Evaluating the performance of waste heat boilers, *Chem. Eng.*, November 16, 1981.
12. *Catalytic Control of VOC Emissions — A Guidebook*, Manufacturers of Emission Controls Association (MECA), 1992.
13. Mink, W.H., Calculator program aids quench-tower design, *Chem. Eng.*, December 3, 1979.
14. National Fire Protection Association Standard NFPA 85B — Standard for the Prevention of Furnace Explosions in Natural Gas-Fired Multiple Burner Boiler-Furnaces, Quincy, MA, 1995.
15. Clark, D.G. and Sylvester, R.W., Ensure process vent collection system safety, *Chem. Eng. Progr.*, January 1996.
16. Howard, W.B., Process safety technology and the responsibility of industry, *Chem. Eng. Progr.*, September 1988.

Bibliography

1. American Petroleum Institute (API) 6th Annual Report,Washington, D.C., May 1998.
2. HPI in brief, *Hydrocarbon Process.*, p. 11, October 1996.
3. Economics of a multimedia approach, *Pollut. Engineering*, p. 42, February 1996.
4. *Journal of Air and Waste Management Association*, p. 119, February 1995.
5. Code Of Federal Regulations (CFR), 40 CFR Part 60, 1997.
6. The Clean Air Act amendments of 1990 – A detailed analysis, *Hazardous Waste Consult.*, p. 4.1, January/February 1991.
7. What to do and when to do it, *Air Pollut. Consult.*, Elsevier Science, p. 4.1, May/June 1997.
8. EPA's gameplan for fighting air toxics, *Enviro. Prot.*, p. 23, October 1998.
9. The new source review reform proposal, *Enviro. Manager*, September 1998.
10. *The Plain English Guide to the Clean Air Act*, EPA 400-k-93-001, Washington, D.C., April 1993.
11. Schedule set for establishing MACT standards, *Air Pollut. Consult.*, p. 2.31, March/April 1994.
12. Understanding the air polution laws that affect CPI plants, *Chem. Eng. Progr.*, p. 30, April 1992.
13. The clean air act ammendments of 1990 – Title I non-attainment, *Hazmat World*, p. 46, October 1991.
14. Dellinger, B.D., Taylor, P.H., and Lee, C.C., Development of thermal stability ranking of hazardous organic compound incinerability, *Enviro. Sci. Technol.*, Vol. 24, March 1990.
15. Morgan, J.L., Hansen, G.M., Whipple, N., and Lee, K.C., Revised model for the prediction of the time-temperature requirements for thermal destruction of dilute organic vapors and its usage for predicting compound destructibility, 75th Annu. Meet. Air Pollution Control Assoc., New Orleans, June 20–25, 1982.
16. Nutcher, P.B. and Lewandowski, D.A., Maximum achievable control technology for NOx emissions from VOC thermal oxidation, AWMA 87th Annu. Meet., Cincinnati, June 19–24, 1994.
17. Vandaveer, F.E. and Segeler, C.G., Combustion of gas, in *Gas Engineers Handbook*, Chapter 5, Section 2, American Gas Assoc., Industrial Press, McGraw Hill, New York, 1965.
18a. *McGraw Hill Dictionary of Chemical Terms*, 1985. 18.
18b. *Guidance on Setting Permit Conditions and Reporting Trial Burn Results*, Vol. 2, Hazardous Waste Incineration Guidance Series, EPA/625/6-89/019, Washington, D.C., January 1989.
19. *Catalytic Control of VOC Emissions – A Guidebook*, Manufacturers of Emission Controls Association (MECA), Washington, D.C., 1992.
20. Chu, W. and Windawi, H., Control VOCs via catalytic oxidation, *Chem. Eng. Progr.*, March 1996.
21. Van Benschoten, D., Pilot test guide VOC control choice, *Environ. Prot.*, October 1993.
22. Heck, R., Farrauto, R., and Durilla, M., Employing metal catalysts for VOC emission control, *Pollut. Eng.*, April 1998.

23. Ciccolella, D. and Holt, W., Systems control air toxics, *Environ. Prot.*, September 1992.

24. Otchy, T.G., First large scale catalytic oxidation system for PTA plant CO and VOC abatement, 85th Annu. Meet. AWMA, June 21–26, 1992.

25. Parker, S.P., Ed., *McGraw-Hill Dictionary of Scientific and Technical Terms*, 4th ed., McGraw Hill, New York, 1989.

26. Reed, J., Ed., *North American Combustion Handbook*, 2nd ed., North American Manufacturing Company, Cleveland, OH, 1978.

27. Code of Federal Regulations, 40 CFR 60.

28. Vandaveer, E. and Segeler, C.G., Combustion of gas, *Gas Engineers Handbook*, Chapter 5, Section 2, Industrial Press, 1965.

29. Crowl, D.A. and Louvar, J.F., *Chemical Process Safety: Fundamentals with Applications*, Prentice Hall, Englewood Cliff, NJ, 1990 .

30. Reed, R.D., *Furnace Operations*, 3rd ed., Gulf Publishing, Houston, TX, 1981.

31. Walas, S.M., *Chemical Process Equipment — Selection and Design*, Butterworths, Stoneham, MA, 1988.

32. Robinson, R.N., *Chemical Engineering Reference Manual*, 4th ed., Professional Publications, Belmont, CA, 1987.

33. Waldern, P.J., Nutcher, P.B., and Lewandowski, D.A., Options for VOC reduction in a regenerative thermal oxidizer (RTO), presented at AWMA Specialty Conf. Emerging Solutions to VOC & Air Toxics Control, San Diego, CA, February 1997.

34. Horie, E., *Ceramic Fiber Insulation Theory and Practice*, Eibun Press, Osaka, Japan.

35. *Handbook of Refractories for Incineration Systems*, Harbison-Walker Refractories, Pittsburgh, PA, 1991.

36. Brosnan, D.A., Crowley, M.S., and Johnson, R.C., CPI drive refractory advances, *Chem. Eng.*, October 1998.

37. Neal, J.E. and Clark, R.S., Saving heat energy in refractory-lined equipment, *Chem. Eng.*, May 1981.

38. Beaulieu, P., Selection criteria for refractory linings of incinerators and acid quench units, 1993 Incineration Conference.

39. Niessen, W.R., *Combustion and Incineration Processes*, Marcel Dekker, New York, 1994.

40. Brunner, C.R., *Incineration Systems*, McGraw Hill, New York, 1991.

41. Reed, R.J., Ed., *North American Combustion Handbook*, Vol. 2, 3rd ed., North American Manufacturing Company, Cleveland, OH, 1997.

42. Damiani, R.A., One stop shopping for heat transfer fluids, *Process Heating*, May/June 1995.

43. Sherman, J., The heat is on, *Chem. Eng.*, November 1991.

44. Cuthbert, J., Choosing the right heat transfer fluid, *Chem. Eng.*, July 1994.

45. Green, R.L. and Morris, R.C., Heat transfer fluids — Too easy to overlook, *Chem. Eng.*, April 1995.

46. Lewandowski, D.A., Economics of heat recovery in the thermal oxidation of wastes, paper read at 86th Annu. Meet. AWMA, Denver, June, 1993.

47. Novak, R.G., Troxler, W.L., and Dehnke, T.H., Recovering energy from hazardous waste incineration, *Chem. Eng.*, March 19, 1984.

48. Ganapathy, V.G., Understanding boiler performance characteristics, *Hydrocarbon Process.*, August 1994.

49. Ganapathy, V.G., Effective use of heat recovery steam generators, *Chem. Eng.*, January 1993.

50. Kiang, Y.H., Predicting dewpoints of acid gases, *Chem. Eng.*, February 9, 1981.

51. Stoa, T.A., Formulas estimate data for dry saturated steam, *Chem. Eng.*, December 10, 1984.
52. Ganapathy, V.G., Evaluating the performance of waste heat boilers, *Chem. Eng.*, November 16, 1981.
53. Balan, G.P., Hariharabaskaran, A.N., and Srinivasan, D., Empirical formulas calculate steam properties quickly, *Chem. Eng.*, January 1991.
54. Dickey, D.S., Practical formulas calculate water properties, *Chem. Eng.*, November 1991.
55. Ganapathy, V.G., Heat recovery boilers: The options, *Chem. Eng. Progr.*, February 1992.
56. Ganapathy, V.G., Win more energy from hot gases, *Chem. Eng.*, March 1990.
57. Parish, M.G., Advantages of heat recovery in air pollution control systems, *Air Pollut. Consult.*, November/December 1991.
58. Burley, J.R., Don't overlook compact heat exchangers, *Chem. Eng.*, August 1991.
59. Guzman, R., Speed up heat exchanger design, *Chem. Eng.*, March 14, 1988.
60. Reynolds, J.P., Dupont, R.R., and Theodore, L.J., *Hazardous Waste Incineration Calculations*, John Wiley & Sons, New York, 1991.
61. Kern, D.Q., *Process Heat Transfer*, McGraw Hill, New York, 1950.
62. Hougen, O.A., Watson, K.M., and Ragatz, R.A., *Chemical Process Principles*, 2nd ed., John Wiley & Sons, New York, 1964.
63. Ganapathy, V.G., *Waste Heat Boiler Deskbook*, Prentice Hall, Englewood Cliffs, NJ, 1991.
64. Bonner, T., Fullenkamp, J., Desai, B., Hughes, T., Kennedy, E., McCormick, R., Peters, J., and Zanders, D., *Hazardous Waste Incineration Engineering*, Noyes Data, Park Ridge, NJ, 1981.
65. Ganapathy, V.G., Understand the basics of packaged steam generators, *Hydrocarbon Process.*, July 1997.
66. Ganapathy, V.G., HRSG temperature profiles guide energy recovery, *Power*, September 1988.
67. Sudnick, J.J., A practical approach to meeting MACT standards with process evaluation, paper read at AWMA Specialty Conf. Emerging Solutions to VOC & Air Toxics Control, Clearwater, FL, February 1996.
68. Gribbon, S.T.J., Regenerative catalytic oxidation, paper read at AWMA Specialty Conf. Emerging Solutions to VOC & Air Toxics Control, Clearwater, FL, February 1996.
69. Seiwert, J.J., High performance thermal and catalytic oxidation systems with regenerative heat recovery, paper read at AWMA Specialty Conf. Emerging Solutions to VOC & Air Toxics Control, Clearwater, FL, February 1996.
70. Klobucar, J.M., Development and testing of improved heat transfer media for regenerative thermal oxidizers in the wood products industry, paper read at AWMA Specialty Conf. Emerging Solutions to VOC & Air Toxics Control, Clearwater, FL, February 1996.
71. Lewandowski, D.A., Nutcher, P.B., and Waldern, P.J., Advantages of twin bed regenerative thermal oxidation technology for VOC emissions reduction, paper read at AWMA Specialty Conf. Emerging Solutions to VOC & Air Toxics Control, Clearwater, FL, February 1996.
72. Matros, Y.S. et al, Conversion of a regenerative oxidizer into catalytic unit, paper read at AWMA Specialty Conf. Emerging Solutions to VOC & Air Toxics Control, San Diego, CA, February 1997.

73. Grzanka, R., Controlling emissions from a black liquor fluidized bed evaporator using a regenerative thermal oxidizer and a prefilter, paper read at AWMA Specialty Conf. Emerging Solutions to VOC & Air Toxics Control, San Diego, CA, February 1997.

74. Seiwert, J.J., Advanced regenerative thermal oxidation (RTO) technology for air toxics control — Selected case histories, paper read at AWMA Specialty Conf. Emerging Solutions to VOC & Air Toxics Control, San Diego, CA, February 1997.

75. Lewandowski, D.A., Nutcher, P.B., and Waldern, P.J., Options for VOC reduction in a regenerative thermal oxidizer (RTO), paper read at AWMA Specialty Conf. Emerging Solutions to VOC & Air Toxics Control, San Diego, CA, February 1997.

76. Nguyen, P.H. and Chen, J.M., Regenerative catalytic oxidation (RCO) catalysts, paper read at AWMA Specialty Conf. Emerging Solutions to VOC & Air Toxics Control, Clearwater, FL, March 1998.

77. De Santis, F. and Biedell, E.L. The evolution of the RTO: 25 years of innovative solutions to VOC control, paper read at AWMA Specialty Conf. Emerging Solutions to VOC & Air Toxics Control, Clearwater, FL, March 1998.

78. Fu, J.C. and Chen, J.M., Rotary regenerative catalytic oxidizer for VOC emission control, paper read at AWMA Specialty Conf. Emerging Solutions to VOC & Air Toxics Control, Clearwater, FL, March 1998.

79. Thompson, W.L., Ruhl, A.C., and Uberio, M., Novel regenerative thermal oxidizer for VOC control, paper read at AWMA Specialty Conf. Emerging Solutions to VOC & Air Toxics Control, Clearwater, FL, March 1998.

80. De Santis, F., RTO continues evolving to meet VOC, HAP control needs, *Air Pollut. Consult.*, January/February 1999.

81. Berger, J., Municipality controls VOCs with cost saving oxidation equipment, *Environ. Technol.*, November/December 1998.

82. Moretti, E., VOC control: Current practices and future trends, *Chem. Eng. Progr.*, July 1993.

83. Nutcher, P.B. and Wheeler, W.H., Chemical and engineering aspects of low NOx concentration, in *Int. Symp. Industrial Process Combustion Technology*, Newport Beach, CA, October 1980.

84. Nutcher, P.B. and Lewandowski, D.A., Control of nitrogen oxides in waste incineration, Environmental Technology Expo, Chicago, February 1992.

85. Nutcher, P.B. and Lewandowski, D.A., ULTRA low NOx design for thermal oxidation of waste gases, AFRC Int. Symp., Tulsa, October 1993.

86. Nutcher, P.B. and Shelton, H., NOx reduction technologies for the oil patch, Pacific Coast Oil Show and Conf., Bakersfield, CA, November 1985.

87. Besnon, R.C. and Hunter, S.C., Evaluation of primary air vitiation for nitric oxide reduction in a rotary cement kiln, EPA/600/S7-86-034, February 1987.

88. Kunz, R.G., Keck, B.R., and Repasky, J.M., Mitigate NOx by steam injection, *Hydrocarbon Process.*, February 1998.

89. Colannino, J.C., Low cost techniques reduce boiler NOx, *Chem. Eng.*, February 1993.

90. Bartok, W. and Sarofim, A.F., *Fossil Fuel Combustion — A Sourcebook*, John Wiley & Sons, New York, 1996.

91. Wendt, J.O., Corley, T.L., and Morcomb, J.T., Effect of fuel sulfur on nitrogen oxide formation in combustion processes, EPA/600/S7-88/007, July 1988.

92. Lewandowski, D.A. and Chang, R.C., Applying equilibrium analysis to low NOx two-stage incinerator design, paper read at AICHE Annu. Meet., Los Angeles, CA, November 1991.

93. Sadakata, M., Fujioka, Y., and Kuni, D., Effects of air preheating on emissions of NO, HCN, and NH_3 from a two-stage combustion, 18th Int. Symp. Combustion, Combustion Institute (1991), Waterloo, Ontario, August 17–22, 1980.

94. Shelton, H. L., Find the right low-NOx solution, *Environ. Eng. World*, November/December 1996.

95. Siddiqi, A.A. and Tenini, J.W., NOx controls in review, *Hydrocarbon Process.*, October 1981.

96. Kunz, R.G., Smith, D.D., Patel, N.M., Thompson, G.P., and Patrick, G.S., Control NOx from furnaces, *Hydrocarbon Process.*, August 1992.

97. Seebold, J.G., Reduce heater NOx in the burner, *Hydrocarbon Process.*, November 1982.

98. Neff, G.C., Reduction of NOx emissions by burner application and operational techniques, *Glass Technol.*, Vol. 31, No. 2, April 1990.

99. Garg, A., Trimming NOx from furnaces, *Chem. Eng.*, November 1992.

100. Kunz, R.G., Smith, D.D., and Adamo, E.M., Predict NOx from gas-fired furnaces, *Hydrocarbon Process.*, November 1996.

101. Shelton, H.L., Reducing process gas NOx, *Chem. Process.*, June 1997.

102. Katzel. J., Controlling boiler emissions, *Plant Eng.*, October 22, 1992.

103. Latham, C. et al., Reburning — An attractive option for NOx reduction, *Air Pollut. Consult.*, November/December 1998.

104. Lewandowski, D.A. and Nutcher, P.B., Control of nitrogen oxides in waste incineration, Environmental Technology Expo, Chicago, February 1992.

105. Lewandowski, D.A. and Nutcher, P.B., Maximum achievable control technology (MACT) for NOx emissions from VOC thermal oxidation, paper read at AWMA Annu. Meet., June 1994.

106. Lewandowski, D.A. and Donley, E.J., Optimized design and operating parameters for minimizing emissions during VOC thermal oxidation, paper read at 88th Annu. Meet. AWMA, San Antonio, TX, 1995.

107. Lewandowski, D.A. and Leaver, G., NOx emissions control techniques in waste gas thermal oxidation, 1st Int. Symp. Incineration and Flue Gas Treatment, Sheffield, England, July 1997.

108. Nutcher, P.B. and Shelton, H.L., NOx reduction technologies for the oil patch, paper read at Pacific Coast Oil Show and Conf., Bakersfield, CA, November 1985.

109. Nutcher, P.B., Forced draft, Low-NOx burners applied to process fired heaters, *Plant/Operations Progress*, Vol. 3, No. 3, July 1984.

110. Wheeler, W. and Nutcher, P.B., Chemical and engineering aspects of low NOx concentration, Int. Symp. Industrial Process Combustion Technology, October 1980.

111. McQuigg, K. and Johnson, B., The effects of operating conditions on emissions from a fume incinerator, Int. Incineration Conf., Knoxville, TN, 1994.

112. Sourcebook: NOx Control Technology Data, EPA/600/2-91-029, Washington, D.C.

113. Bai, H. and Yu, H., Ammonia injection: A new approach for incinerator emissions control, 86th Annu. Meet. AWMA, Denver, June 1993.

114. Staudt, J.E., Confuorto, N., Grisko, S.E., and Zinsky, L., NOx reduction using the NOxOUT process in an industrial boiler burning fiberfuel and other fuel, ICAC Forum '93, Washington, D.C.

115. Brouwer, J., Heap, M.P., Pershing, D.W., and Smith, P.J., A model for prediction of selective non-catalytic reduction of nitrogen oxides by ammonia, urea, and cyanuric acid with mixing limitations in the presence of CO, 26th Int. Symp. Combustion.

116. Caton, J.A. and Siebers, D.L., Comparison of nitric oxide removal by cyanuric acid and by ammonia, *Combust. Sci. Technol.*, Vol. 65, 1989.

117. Pickens, R.D., Add-on control techniques for nitrogen oxide emissions during municipal waste combustion, *J. Hazardous Materials*, No. 47, 1996.

118. White paper — Selective non-catalytic reduction (SNCR) for controlling NOx emissions, Institute of Clean Air Companies (ICAC), Washington, D.C, October 1997.

119. Jodal, M., Lauridsen, T.L., and Dam-Johansen, K. NOx removal on a coal-fired utility boiler by selective non-catalytic reduction, *Environ. Progr.*, Vol. 11, No. 4, November 1992.

120. Sellakumar, K.M., Isaksson, J., and Tiensuu, J., Process performance of Ahlstrom pyroflow PCFB pilot plant, paper presented at 1993 Int. Conf. Fluidized Bed Combustion, San Diego, CA, May 9–13, 1993.

121. Lewandowski, D.A. and Hamlette, B.J., Performance parameters for post-combustion NOx control using ammonia, AWMA Specialty Conf. Emerging Solutions to VOC & Air Toxics Control, Clearwater, FL, March 1998.

122. Jones, D.G. et al., Two-stage DeNOx process test data for 330 TPD MSW incineration plant, paper presented at 82nd APCA Annu. Meet. and Expo, Anaheim, CA, June 1989.

123. Lange, H.B. and DeWitt, S.L., Plume visibility related to ammonia injection for NOx control — A case history, NOx Control VII Conf., Council of Industrial Boiler Owners, Chicago, IL, May 1994.

124. White Paper — Selective catalytic reduction (SCR) for control of NOx emissions, Institute of Clean Air Companies (ICAC), Washington, D.C, November 1997.

125. Siddiqi, A.A. and Tenini, J.W., NOx controls in review, *Hydrocarbon Process.*, October 1981.

126. Cho, S.M., Properly apply selective catalytic reduction for NOx removal, *Chem. Eng. Progr.*, January 1994.

127. Selective catalytic reduction makes inroads as NOx control method, *Air Pollut. Consult.*, May/June 1995.

128. Czarnecki, L., Put a lid on NOx emissions, *Pollut. Eng.*, November 1994.

129. Heck, R.M., Chen, J.M., and Speronello, B.K., Operating characteristics and commercial operating experience with high temperature SCR NOx catalyst, *Environ. Progr.*, Vol. 13, No. 4, November 1994.

130. *Evaluation of NOx Removal Technologies*, Vol. 1, *Selective Catalytic Reduction*, Revision 1, U.S. Department of Energy, Pittsburgh Energy Technology Center, February 1994.

131. Brown, C., Pick the best acid-gas emission controls for your plant, *Chem. Eng. Progr.*, October 1998.

132. Pan, Y.S., Review of flue gas desulfurization technologies, *Air Pollut. Consult.*, May/June 1992.

133. Brady, J.D., Combat incinerator off-gas corrosion, Part 1, *Chem. Eng. Progr.*, January 1994.

134. Buonicore, A.J., Experience with air pollution control equipment on hazardous waste incinerators, 83rd Annu. Meet. AWMA, Pittsburgh, June 1990.

135. Brady, J.D., Dry effluent — Wet scrubbing system for waste incinerators, 81st Annu. Meet. Air Pollution Control Association, June 1988.

136. Getz, N.P., Amos, C.K., and Siebert, P.C., Air pollution control systems and technologies for waste-to-energy facilities, *Energy Eng.*, Vol. 88, No. 6, 1991.

137. Bacon, G.H., Ramon, L., and Liang, K.Y., Control particulate and metal HAPs, *Chem. Eng. Progr.*, December 1997.

138. McInnes, R., Jameson, K., and Austin, D., Scrubbing toxic inorganics, *Chem. Eng.*, February 1992.

139. Bendig, L., Wet scrubbers: Match the spray nozzle to the operation, *Environ. Eng. World*, March/April 1995.
140. Croom, M.L., Effective selection of filter dust collectors, *Chem. Eng.*, July 1993.
141. Nudo, L., Capturing heavy metals, *Pollut. Eng.*, September 1993.
142. Hulswitt, C.E., Adiabatic and falling film absorption, *Chem. Eng. Progr.*, February 1973.
143. Mink, W.H., Calculator program aids quench-tower design, *Chem. Eng.*, December 3, 1979.
144. National Electrical Code, ANSI/NFPA 70, Article 500.
145. Definitions and information pertaining to electrical instruments in hazardous (classified) locations, Standard ISA-S12.1-1991, Instrument Society of America.
146. Fume incinerators, Standard 6–11, Factory Mutual (FM).
147. Clark, D.G. and Sylvester, R.W., Ensure process vent collection system safety, *Chem. Eng. Progr.*, January 1996.
148. Standard on explosion prevention systems, NFPA, Quincy, MA, 1986.
149. Lewandowski, D.A. and Waldern, P.J., Design and operating parameters for thermal oxidation of volatile organic compounds, presented at Incineration Conf., Seattle, 1995.
150. Lewis, B. and Von Elbe, G., *Combustion, Flames, and Explosion of Gases*, 2nd ed., Academic Press, New York, 1961.
151. Mendoza, V.A., Smolensky, V.G., and Straitz, J.F., Don't detonate — Arrest that flame, *Chem. Eng.*, May 1996.
152. Bishop, K. and Knittel, T., Do you have the right flame arrestor?, *Hydrocarbon Process.*, February 1993.
153. Mendoza, V.A., Smolensky, V.G., and Straitz, J.F., Do your flame arrestors provide adequate protection?, *Hydrocarbon Process.*, October 1998.
154. Laust, P.B. and Johnstone, D.W., Use nitrogen to boost plant safety and productivity, *Chem. Eng.*, June 1994.
155. Gooding, C.H., Estimate flash point and lower explosive limit, *Chem. Eng.*, December 12, 1983.
156. Shelton, H.L., Estimating the lower explosive limits of waste vapors, *Environ. Eng. World*, May/June 1995.
157. Howard, W.B., Process safety technology and the responsibility of industry, *Chem. Eng. Progr.*, September 1988.
158. Howard, W.B., Flame arrestors and flashback preventers, presented at 16th Annu. Loss Prevention Symp., Anaheim, CA, June 7–10, 1982.
159. Standard for the Prevention of Furnace Explosions/Implosions in Multiple Burner Boilers, 1995 ed., NFPA 8502, Quincy, MA.
160. Jensen, J.H., Combustion safeguards for gas and oil fired furnaces, *Chem. Eng. Progr.*, October 1978.

Index